T0284720

THE MASSEY LECTURES SERIES

The Massey Lectures are co-sponsored by CBC Radio, House of Anansi Press, and Massey College in the University of Toronto. The series was created in honour of the Right Honourable Vincent Massey, former Governor General of Canada, and was inaugurated in 1961 to provide a forum on radio where major contemporary thinkers could address important issues of our time.

This book comprises the 1970 Massey Lectures, "Therefore Choose Life," broadcast in November 1970 for Radio Canada International. The producer of the series was Lewis Auerbach and the lectures were transcribed by Ann Coombe for the CBC, with corrections by George Wald and Elijah Wald.

GEORGE WALD

George Wald was born in New York City in 1906, the youngest son of Jewish immigrants from Germany and Poland. The first member of his family to go to college, he became an award-winning biologist, teaching at Harvard University for forty-three years. In 1966, *Time* magazine listed him in a cover story as "one of the ten best teachers in the country." Wald's long research career began with his discovery of vitamin A in the eye. His further explorations of the chemistry and physiology of vision led to a 1967 Nobel Prize in Physiology or Medicine, shared with Haldan Keffer Hartline and Ragnar Granit. He was elected to the National Academy of Sciences in 1950, the American Philosophical Society in 1958, and from 1963 to 1964 he was a Guggenheim Fellow, spending the year at Cambridge University. He also received honorary degrees from the University of Berne, Yale University, Wesleyan University, New York University, McGill University, Clark University, and Amherst College. Wald spoke out on many political and social issues, and his fame as a Nobel laureate brought national and international attention to his views. He was a vocal opponent of the Vietnam War and the nuclear arms race, served on the Russell Tribunal on Human Rights, and worked for social justice in a broad range of national and international settings. In 1997, Wald died at his home in Cambridge, Massachusetts, at the age of ninety.

THEREFORE CHOOSE LIFE

GEORGE WALD

ANANSI

Copyright © 1970, 2017 Elijah Wald and the Canadian Broadcasting Corporation
Foreword © 2017 Elijah Wald
Introduction © 2017 Lewis Auerbach
"An Interview with George Wald" © 1970, 2017 Elijah Wald and Lewis Auerbach

Published in Canada in 2017 and the USA in 2017 by House of Anansi Press Inc.
www.houseofanansi.com

All rights reserved. No part of this publication may be reproduced or transmitted
in any form or by any means, electronic or mechanical, including photocopying,
recording, or any information storage and retrieval system, without permission
in writing from the publisher.

"Do Not Go Gentle Into That Good Night" (four-line excerpt appearing on page 10)
by Dylan Thomas, from THE POEMS OF DYLAN THOMAS, copyright © 1952 by Dylan
Thomas. Reprinted by permission of New Directions Publishing Corp.

House of Anansi Press is committed to protecting our natural environment.
As part of our efforts, the interior of this book is printed on paper that contains
100% post-consumer recycled fibres, is acid-free, and is processed chlorine-free.

21 20 19 18 17   1 2 3 4 5

Library and Archives Canada Cataloguing in Publication

Wald, George, 1906-1997, author
Therefore choose life : the found Massey lectures
/George Wald.

(CBC Massey lectures)
Includes index.
Issued also in electronic formats.
ISBN 978-1-4870-0320-3 (softcover).—ISBN 978-1-4870-0321-0
(EPUB).—ISBN 978-1-4870-0322-7 (Kindle)

1. Science—Social aspects. 2. Science—Moral and ethical
aspects. 3. Science—Philosophy. 4. Life. 5. Evolution. 6. Bioethics.
7. Religion and science. I. Title. II. Series: CBC Massey lectures
series

Q175.5.W34 2017   303.48'3   C2017-902935-5
                         C2017-902936-3

Library of Congress Control Number: 2017940462
U.S. ISBN 978-1-4870-0338-8

Cover design: Alysia Shewchuk
Text design: Ingrid Paulson
Typesetting: Alysia Shewchuk

*We acknowledge for their financial support of our publishing program
the Canada Council for the Arts, the Ontario Arts Council, and the Government of
Canada through the Canada Book Fund.*

Printed and bound in Canada

CONTENTS

Foreword   *ix*
Introduction   *xvii*
A Note from the Publisher   *xxiii*

Chapter One: One with the Universe   1

Chapter Two: The Origins of Life   15

Chapter Three: The Origins of Man   33

Chapter Four: The Origins of Death   45

Chapter Five: Answers   63

Chapter Six: A Question of Meaning   79

An Interview with George Wald   *93*
Index   *125*

I call heaven and earth to record this day against you, that I have set before you life and death, blessing and cursing: therefore choose life, that both thou and thy seed may live.

—King James Bible, Deuteronomy 30:19

FOREWORD

MY FATHER, GEORGE WALD, lived two public lives, first as a scientist and then as a social activist. The Massey Lectures, delivered as he was making the transition between those lives, provides a fine summation of the first and introduction to the second — and, most significantly, shows how the science served as a foundation for the activism. He was sixty-four years old in 1970 and had devoted most of his energies over the previous half-century to researching and teaching biology. Inspired by Sinclair Lewis's novel *Arrowsmith*, he approached biology as a pure quest for knowledge, and was fortunate to live in a period when there was still funding for that kind of work, with relatively little pressure to produce medical breakthroughs or commercial products.

He was also fortunate to have made the discovery of vitamin A in the eye during his first year of post-graduate work in Germany in the early 1930s. This discovery set him on the path he would pursue throughout his scientific career. With a series of associates — including my mother, Ruth Hubbard — he worked out much of the chemistry of visual pigment and the ways eyes react to light. He found this research exciting and profoundly meaningful, saying, "A scientist is, in a way, an artist — I have always felt that I was developing a field, and that this was something like painting a picture."

That part of his life culminated in a 1967 Nobel Prize for Physiology or Medicine, which marked the beginning of the end of his scientific career and his transition into social and political activism. That transition was sparked by his experiences as a teacher, a career that meant as much to him as his research work, and which for him was always a reciprocal process. He sometimes said, with pride, that he was the only science professor at Harvard who had ever asked to teach a freshman introductory course. He insisted upon teaching first-year biology courses not only because he liked to introduce students to his field and get them excited about it, but because he loved talking with young people — it was not unusual, at a party, to find him in a corner with a five-year-old, earnestly discussing cosmology or natural history, and coming away energized, saying he could never have had as stimulating a conversation with an adult.

In the 1960s, when student political activism exploded in reaction to the civil rights movement and the war in Vietnam, he was cheered and inspired by the new spirit of radicalism and social commitment. That attitude was unusual for a professor of his age and prominence — as I recall, he was one of only six Harvard faculty members to support the mass student strike in 1969. In March of that same year, he gave a more or less impromptu speech at a teach-in against the Vietnam War at the Massachusetts Institute of Technology, which he mentions in these lectures. Titled "A Generation in Search of a Future," it was immediately recognized as a major statement, reprinted across the country, translated into forty languages, and released as a record album. Virtually overnight, my father became a leading peace activist, travelling around the United States and abroad.

The Massey Lectures captures him in that moment of transition, showing the deep foundations of his beliefs and his effort to connect science with social and political action. The first two chapters revisit themes from his first great popular science lecture, "The Origin of Life," and the fourth is largely drawn from its sequel, "The Origin of Death" — a surprisingly optimistic meditation on the role of death in nurturing life. The other three chapters seek in various ways to apply the lessons of those scientific explorations to questions and issues in the world around him, a process he would pursue for the rest of his life.

That process was always evolving: for my father, the great virtue of science was its open-endedness — not that it provides correct answers, but that it treats all answers, right or wrong, as steps in a journey to acquire knowledge. Recent climate change deniers have cited him as an example of scientific fallibility and alarmism for warning in 1970 that civilization would end in fifteen or thirty years unless dramatic actions were taken. His response would be that in an experiment the size of a planet, there is nothing surprising about predictions being off by decades, and so far the trends that worried him in 1970 have only grown more serious and dangerous. (The critics' obliviousness to those dangers recalls the old joke about a man who falls off the Empire State Building and, as he falls past the thirty-first floor, is heard to say, "So far, so good.")

While most of the concerns expressed in these lectures remain current a half-century later, I am struck by some archaisms. The 1970s saw the rise of the modern feminist movement, and within a few years my father would not so casually have used "man" as a synonym for "human." His enthusiasm about the potential of nuclear power is particularly surprising, since he became an active opponent of that industry by the 1980s — though he continued to think there were some possibilities for fission, he concluded that fusion reactors were simply too dangerous. He also became more wary of framing overpopulation as a central issue; already in these lectures he was careful to note that the main stress on the Earth's resources came from the developed world, and he became increasingly convinced that overpopulation should be dealt with as a symptom rather than as a cause of that stress. The important thing was to improve the standards and quality of life for people everywhere, and in particular to focus on nurturing the most basic human values, shaping "a better world for children."

I would add a note on the organization my father discusses in the interview following the last lecture, which he was then forming and which shortly became the Labor-University Alliance. He had always taken pride in his background in working-class Brooklyn, and hoped to bridge the growing gaps between students and workers, forging alliances between the core constituencies of the old and new Lefts. For various reasons, the organization never quite took off and lasted only a couple of years — but it was an important effort, and some connections formed in that period had longer-lasting offshoots.

There is far more to be said, but fortunately these lectures have survived and I'm glad to have my father speak for himself. As a final word of introduction, this is what he often told students and interviewers:

> One of the most important sources of human happiness is to find an unachievable objective. That sounds strange, but in this life there are many things you want — you want to find someone you can love, you want to build a home, you want to have children. And many of these things you will do — but somehow the finding will never quite come up to the dreams that went into it. So it's important to find one goal that never stops being a goal, where you can have little victories,

but they are just incidents in that bigger thing.
Science fulfills such a role...one need never fear
that one will come to the end of the enterprise.

And, he would add:

When I was a young student I used to be told that
a scientist always asks how, but never why. I have
come to think that a degraded view of the scientist. To be sure we ask how again and again, but
if we have had the good fortune to be answered,
there comes a time to ask why.

Elijah Wald
Philadelphia, Pennsylvania
March 2017

INTRODUCTION

IN THE SPRING OF 1970, I convinced my colleagues at
CBC *Ideas* that George Wald give the 1970 Massey
Lectures. I suggested Wald because he was a Nobel
Prize–winning scientist who knew how to make biology understandable, fascinating, and relevant. His
influence extended from his scientific accomplish-
ments on the physiology of vision, to his extraordinary impact on his students, to his capacity to
express the importance of knowledge over ignorance.

Wald was also widely admired for his coura-
geous moral positions. He publicly supported an
end to U.S. engagement in the war in Vietnam.
He opposed university involvement in war-related
work and supported campaigns for birth control.
All this connected with his deep reverence for
understanding the origins of life. As he put it in his

first lecture, "One With the Universe," "It is a grand story, the grandest story I know, the most awesome, the most beautiful, the most meaningful."

Every Harvard student, I among them, knew about Wald's hugely popular introductory course in biology. The *Harvard Crimson* reported the course "turned more scientists into poets, and more poets into scientists, than any course ever taught on this campus." My classmate, Governor General David Johnston, took the course, and it influenced him, as it did many others, profoundly. "(T)he light went on," he told the *Varsity*, "permitting you to triangulate and look at problems from a different angle. You begin to appreciate the importance of bringing other points of view and using those other emphases." Johnston told me that this recognition of the impact of science continued throughout his life and was especially important to his many writings about legal issues with respect to computers, communications, and cyberspace.

My invitation to Dr. Wald had emphasized that he would be the first natural scientist to receive the honour of delivering the Massey Lectures. "This is not due to any anti-scientific bias on our part — quite the contrary," I said. "At a time when there is so much concern about individual *survival*, (psychic, physical, etc.) and also about the survival of the human race upon the planet, there are a number of people who are looking ever more intensely to science for solutions."

Indeed, 1970 was a momentous year. Musically it began with the release of "Bridge Over Troubled Water" — a song whose spirit is in the air again — and ended with the breakup of the Beatles and the release of "Let It Be." Protest was in the air. Even though the '60s were over, the Vietnam War raged on and the National Guard would shoot four students at a protest demonstration at Kent State University. In Canada, draft dodgers were still arriving, and even more significant was the October Crisis, which saw FLQ separatists kidnapping a British trade commissioner, the federal response of imposing the War Measures Act, and the murder of Quebec labour minister Pierre Laporte.

Wald responded to my invitation positively and almost immediately, inviting me to record the talks in his study at Woods Hole, Massachusetts, where he spent the summers teaching and doing research at the Marine Biological Laboratory. These were the days before the Massey Lectures were delivered in front of live audiences across Canada. After we finished some of the lectures, he insisted that we record post-talk interviews, while he piloted his small sailboat. I can still remember hoping the tape recorder wouldn't get wet.

The lectures have several profound themes, among them the value of life, the importance of death, the long-term negative consequences of inequality and pollution, the role of religion, and the importance of diversity. Wald spoke of these issues with respect, deep knowledge, and even humour.

Concluding one of the lectures with a biblical quote, Wald urged listeners to care about our future and the future of others: "Nowadays there is a beautiful passage from Deuteronomy always running through my thoughts. It's chapter 30, verse 19. Let me say it for you. 'I have set before you life and death, blessing and curse; therefore choose life, that you and your descendants may live.'"

The programs were a huge success. One reason was Wald's insistence that he not read from a script, but rather speak from the heart. This was unusual, but as radio producers first and foremost, we thought it was a reasonable request, since Wald promised he would edit the transcripts for publication. Besides, it was the only way he felt comfortable delivering lectures, whether to Harvard students or to a broader public. As his son Elijah notes, he typically "spoke without a written text or notes…using only a few cards with key phrases written on them."

After the programs were broadcast, and despite our requests and encouragement, it became evident that he would never produce the promised manuscript for a book. Indeed, this was not the only time he had trouble producing a book — he had a great deal of trouble finishing a manuscript of a biology textbook as well.

But I never completely lost hope. Then, a few years ago, some "lost" Massey Lectures, including those of Martin Luther King Jr., were "found" and republished. Perhaps *Therefore Choose Life* would be one of them? But it was not to be. There was, as far as anyone knew, no manuscript. No one knew where the transcript was anymore, and Dr. Wald had died in 1997.

In 2016, fate intervened. On a whim one July morning, I googled "George Wald," "CBC," and "Therefore Choose Life," and bingo! Halfway down the page was a link to the George Wald papers at the Harvard Archives. In Harvard's online catalogue of his papers, there were references to transcripts and correspondence relating to the lectures. This was amazing — after all these years, it would be possible to publish the book! All I had to do was get myself to Cambridge to see them.

When I arrived, the archivists had dug out the relevant boxes from Dr. Wald's papers. Within minutes, I found his edited transcripts. And since the programs had been broadcast on Radio Canada International, there were also letters from listeners all over the world. Many asked Wald when the book would be published. He replied with a form letter that began: "Thank you. As soon as I can find the time, I will edit the transcript of my Massey Lectures."

I wrote to Greg Kelly, the executive producer of *Ideas*, to tell him a truly lost Massey Lectures had now been found and that nearly a half-century later, the lectures are still incredibly relevant, even more so today than they might have been just a few years ago. Greg was delighted: "The Masseys have this astonishing afterburn, reaching listeners and readers well beyond the year they were produced."

House of Anansi Press agreed that *Therefore Choose Life* should be "found" at last.

The rest is history.

Lewis Auerbach
Ottawa, Ontario
March 13, 2017

A NOTE FROM THE PUBLISHER

Therefore Choose Life was originally broadcast in 1970. Throughout the lectures, the word "man" is commonly used as a synonym for all human beings. Wald had stopped using "man" in this way by the 1980s, and in later decades would also have said Inuit rather than Eskimo. We have left these terms as they were in the original, but they do not reflect either his or the publisher's later preferences.

THEREFORE
CHOOSE LIFE

ONE

ONE WITH THE UNIVERSE

IN THE BOSTON MUSEUM OF FINE ARTS, there is a large wood carving by Paul Gauguin, from his Tahitian period. It is a brooding, mystical piece and, up in the upper left-hand corner, Gauguin carved three phrases. In French they are: *"D'où venons-nous? Que sommes-nous? Où allons-nous?"* In English: "Whence do we come? What are we? Whither are we going?" All men, everywhere, have asked those same questions: whence we come, what kind of thing we are, and at least some intimation of what may become of us. Seeking answers to these questions, men have followed many paths. I hope I may be forgiven for believing that science offers what is, perhaps, the surest of those paths.

We have special need now for answers to those questions. Our society is in a crisis of conviction, of mission, of commitment — a kind of worldwide identity crisis. Technology having obliterated distance, man needs more than ever before to become a community; unless we can achieve some commonly accepted sense of human needs and goals, we're lost. That is the kind of thing I shall be talking about in these chapters. I shall be asking the question: From what base can a scientist make moral and political judgements? I would like to examine that base, my base. Perhaps it can become yours. What I am looking for is some sort of context that can serve as a guide to decision and action.

In a sense, this is my religion, the entirely secular religion of one scientist. It contains no supernatural elements. Nature is enough for me — enough of awe, enough of beauty, enough of faith and reason. I should seek out the supernatural only if I felt that we had exhausted nature; but we haven't, and we never will.

I would like to begin by sorting out some basic ideas. We need to know what we are talking about. First, man has been engaged, ever since we have known him, in an unending struggle to know. I think that constant search for knowledge is epitomized in science. Science is an attempt to understand all reality. Reality covers a very broad province, not only such relatively simple things as stones falling and the structures of atomic nuclei, but much more complicated things, such as poets writing sonnets, people weeping, people praying. I think that science will never understand some of those more complicated things, but we'll keep on trying. The point of the whole enterprise is to achieve *understanding*. Facts are only the raw material of science.

Some time ago I read for the first time, though not the last, Hermann Hesse's *Siddhartha*. I came out of that first reading with a wonderful sentence — the sentence isn't in Hermann Hesse's book, but all the parts are there and that's how I found it. It went: "One can gain knowledge from words, but wisdom only from things." I think that's what science is about: it's a deep-seated attempt to extract the wisdom from things.

As such, that deep and consistent attempt to understand reality is what makes science altogether good, as our culture interprets the good. There can be no such thing as bad science. Any other view would be a plea for ignorance, and there can be no possible quarrel with science that ignorance can improve.

Another entirely different enterprise is the application of science to useful ends: technology. I've just finished saying that science is altogether good, but I would never dream of saying that about technology. Technology is for use and in any properly conducted society every enterprise in technology, new and old, should be under constant review and judgement in terms of the needs, goals, and aspirations of that society.

One of the troubles with our present society is that we tend to regard all technology without question as progress — sometimes the more unpleasant aspects of technology as aspects of fate — and that's altogether wrong. I used to ask sometimes, when it was still a rare question: "Should one do everything one can?" The usual answer was: "Why yes, of course. Of course one does everything one can. One travels as far and as rapidly, makes as big a bomb, and does all those other things one can do, as soon as one becomes able to." But the proper answer is: "Of course not." Among all those things that can be done, decisions must be made as to which to do and which not to do, and they must be made in terms of essential human social needs.

Who is to make those decisions? Well, another trouble with our present society is that those decisions are being made almost entirely by the producers of technology, by those who see in technology opportunities for wealth or power or status. One should listen to all that such interested parties have to say, but the final decision should be made not by the producers of technology, but by those who will have to live with the products. So the position is: *know* all you can, but *do* only what seems socially useful and beneficial.

There is another dichotomy that causes equal confusion and equal trouble. It's the distinction between *creation* and *production*. Creation, again, is altogether good, as our culture interprets the good; but production, again, is for use, and in any properly conducted society, all production should be under constant review and judgement in terms of human and social needs and goals.

One speaks a great deal these days of the alienation of man from his world, and sometimes one ascribes that alienation to science. But I think that's a misunderstanding. Science, properly understood, can't alienate. Its entire point and purpose is to make man more at home in his universe. There was a time, about a century ago, when scientists, anxious to sweep aside the accumulated rubbish of centuries of tradition, undercut man's view of his place

in life and in the universe and substituted nothing
for what it was taking away. But by the beginning
of this century, having completed that task, science
began to put the world back together again, and by
now it has achieved a remarkably unified view of
the kind of universe we're in, of the place of life in
it, and of the place of man in life.

What is that unity? Well, we know now that
we live in a historical universe, one in which not
only living organisms, but stars and galaxies, are
born, come to maturity, grow old, and die. That
universe is made of four kinds of elementary par-
ticle: protons, neutrons, electrons, and photons,
which are particles of radiation. One could begin
such a universe with just neutrons, for within a
half-life of something like thirteen minutes those
neutrons would have disintegrated partly into protons, electrons, and radiation. One could begin such
a universe with just light, radiation — "Let there be
light!" — for photons, the particles of light, are, as
we shall shortly see, interconvertible with those
other particles.

If not the whole universe, then surely large parts
of it began as a gas, a so-called plasma of such elementary particles filling large sections of space.
Then, here and there in that gas of elementary

particles, just by chance, an eddy formed, a little special knot of material, and this began pulling in the particles about it through ordinary forces of gravitation. So it began to grow, and as it grew, it pulled in more and more particles. The more of it there was, the better it was at pulling in the particles from the space about it. So we began to have a condensing mass of elementary particles. As such a mass, as any mass, condenses, it heats up; and when the temperature in its deep interior reached about five million degrees, something new began to happen.

That new thing was the so-called burning—it isn't really burning, but it's called that—of hydrogen. One of those elementary particles, the protons—a proton is the nucleus of a hydrogen atom, so one speaks of protons frequently as hydrogen—those protons, those hydrogen nuclei, began to condense to form helium nuclei: four hydrogen nuclei to form one helium nucleus. Four hydrogen nuclei, each of mass 1, to form a helium nucleus of mass about 4. But in this transaction, a little bit of mass is lost. A helium nucleus has a slightly smaller mass than four protons, and that little bit of mass is converted into radiation, according to Einstein's famous formula, $E = mc_2$, in which E is the energy of that radiation, m is that little bit of mass, and c is a very large number, the speed of light: 186,000 miles a second or $3 \times 10_{10}$ centimetres per second — an enormous number. You square a number that big and you get a fantastically big number, and multiply even a tiny bit of mass by that big a number and you have an enormous outpouring of radiation. That radiation begins to be poured out in the deep interior of what had been a collapsing mass of elementary particles and it backs up the further collapse — it brings this mass of material into a kind of uneasy steady state. What I have just described is the birth of a star. That star has now entered its period of maturity; it's, as we say now, on the *main sequence*. Our sun achieved that state — it went on the main sequence — something like six billion years ago. It's an ordinary, run-of-the-mill, middle-aged star. It has about another six billion years to run.

A star, any star, lives on this process, on this conversion of hydrogen to helium. Then, inevitably, the time comes for every star when it begins to run out of hydrogen. As that happens, it produces less energy so it begins to contract again, and as it contracts it begins to heat up again. And when the temperature in the deep interior reaches about one hundred million degrees, something new happens. This time it is the so-called burning of helium: those helium nuclei, each of a mass about 4, begin to combine with one another. It's all simple arithmetic. Two helium nuclei combine: four and four make eight. That's beryllium eight, ^{8}Be, an element so unstable that it disintegrates within so small a fraction of a second that it has never been measured. But at these enormous temperatures and pressures there are always a few beryllium nuclei, and here and there one of them captures another helium nucleus: eight and four make twelve, and what's twelve? That's *carbon*. That's where carbon comes from in our universe. And when you've got some carbon, that carbon twelve can capture another helium nucleus: twelve and four make sixteen, and what's sixteen? That's *oxygen*. That's where oxygen comes from in our universe. The carbon has another trick: it can begin to pick up protons, those hydrogen nuclei, and carbon twelve plus two protons makes fourteen. And what's fourteen? That's *nitrogen*. These new reactions, these new nuclear processes, produce a tremendous new outpouring of energy in the deep interior of that star, enough not only to stop the further collapse, but to puff it up to enormous size. It becomes a red giant: gigantic because it's puffed up, and red because it's cooled off somewhat in its outer layers. A red giant is a dying star. With them it is as with Dylan Thomas's father:

> Do not go gentle into that good night,
> Old age should burn and rave at close of day;
> Rage, rage against the dying of the light.

That's the way it is with red giants. They're in a delicate condition. They're always boiling off a lot of material from their surfaces into outer space. Now and then they send a great streamer of material roaring off into space: a flare. Now and then they threaten to blow up entirely: a nova. Now and then they do blow up: a supernova. In all these ways, the stuff of which red giants are made is returned to space to become part of the enormous masses of gases and dust which fill all of space. It has been estimated that as much as half the mass of our universe is in the form of gases and dust.

Then, here and there in the gases and dust, just by chance, a new eddy forms, a new knot of material, and once again it begins to pull in more material through gravitation, and a new star is born. But these later-generation stars, unlike the first generation that was made exclusively of hydrogen and helium, contain carbon, nitrogen, and oxygen — and we know that our sun is such a later-generation star because we're here . . . *because we are here*. We, like all the other living creatures we know, are made of just those four elements that I have spoken of: hydrogen, carbon, nitrogen, and oxygen. There are ninety-two natural elements, but 99 percent of the living substance of all living organisms we know are made of those four. And I think, though that is another story, that it has to be that way — that wherever life arises anywhere in the universe, it's made primarily of those four elements.

I find it deeply moving to realize that stars have to die so that organisms may live. It's all part of the profound fitness of things in our universe. You see, all the life on Earth runs ultimately on sunlight. The source of that sunlight: nuclear reactions among the four elements that I have been speaking of. The composition of the life: those same four elements bound together in molecules. The stars and the organisms are both living on processes involving the same four elements; the stars on the nuclear reactions of those elements, the organisms on the whole atoms. The stars are too hot for the atomic nuclei within them to draw the electrons about themselves in orderly ways. That happens only in the cooler places in the universe — the planets — and there, where the atomic nuclei can draw the electrons about themselves in orderly ways so as to make whole atoms, the electrons can interact with one another and so achieve molecules.

Molecules are a great new thing in the universe. Until molecules have appeared, nothing has a shape or size or even a determinable location and motion. Until the molecules appear, one is still in the universe of Heisenberg's Indeterminacy or Uncertainty Principle. The elementary particles and atoms are too small to have shapes or sizes or, as I say, even determinable positions and motions. All of those things come into the universe with molecules.

As soon as our planet, Earth, formed some four and a half billion years ago, the molecules began to form on it. And after they had accumulated over some ages of time, those molecules eventually led to life. That is what I shall be speaking of next time. It is a grand story, the grandest story I know — the most awesome, the most beautiful, the most meaningful.

I should like to say one last thing: we have been told so often, and on such tremendous authority—that of Plato and of Kant—as to take it for granted, that the *Ding an sich*, the essence of things, the inner reality, must remain forever hidden from us, that we stand outside our universe like children before a shop window with their noses pressed against the glass, able to look in but unable to enter. It isn't so. We see our universe not from outside, but from inside. We are one with it: its substance, our substance; its history, our history. From that realization, I think we can take some assurance that what we see is real.

TWO

THE ORIGINS OF LIFE

WE LIVE IN A historical universe, and we're talking out some of its history. Why are we doing that? Well, because in knowing something of that history, one gains great assurance in a world that's a little adrift. There is great assurance in realizing that we live in an understandable universe, one that makes sense, one in which things fit together.

Molecules are a great new thing when they appear in the history of the universe. Molecules are a great thing to me. I have lived most of my scientific life with molecules, and they're good company. I tell my students, "Try to get to be intuitive about molecules. Try to get so that when you have some problem involving molecules, you ask yourself, 'What would I do if I were that molecule?' Then all the answers are likely to come out right." I tell my students, "Try to feel like a molecule; and if you work really hard, some day you may get to feel like a *big* molecule."

What we talked about last time involves the extraordinary fitness of things in our universe. The elements that generate the light of the stars also generate life on the planets around those stars, and then that life comes to live on the light.

So that brings us to the question: How did life arise upon the Earth? It's a big question, and not the kind science typically addresses. Science ordinarily has very little to do with beginning and endings. Its business is with things that happen again and again, so we can observe them again and again, and learn their properties. Its business is with the order of nature. A unique event has no place in that order; it smacks always of the unnatural or supernatural, of magic or miracle. Yet men have always insisted upon asking the ultimate questions. In our own tradition, we have an account of the origin of life in the opening paragraphs of the Book of Genesis. There we are told that, beginning on the third day, God began to create life: first, the green plants (fortunately on the fourth day he made the sun), then the fishes of the sea and the birds of the air, then the creatures that creep upon the land, and so onward and upward through the creation until its last and most magnificent work — woman.

That's one story of the way in which life began on the Earth: as an act of supernatural creation. It is a story that's deeply ingrained in our tradition and precious to many persons, and it is thought of as representing a religious view of that event. But in a sense that story makes life a technological product, albeit one produced by God. To me, there is a much grander story to be told, a quite differ- ent view, that says that in this historical universe, given enough time, non-living matter "spontane- ously" brings forth life. It does so inevitably, so that life has a firm place in the order of nature, in the physics of an evolving universe.

It's rather curious that this view of the spontane- ous generation of life from the non-living doesn't, in fact, break with the beautiful story in Gene- sis. For what is told to us in those opening para- graphs of Genesis is not exactly that God made life, but rather that he ordered the earth and waters to "bring forth" life. With such orders having once been given to the earth and waters and never, so far as we know, rescinded, one view is that the earth and waters are free forever after to bring forth life, which is exactly what we mean by spontaneous generation.

So, what does one need to make a living cell, to make a living organism? Well, first and foremost, one needs water. Even the more solid and tougher parts of living organisms are made mostly of water. Animal muscle — a steak, for example — is about four-fifths water, and the softer parts of organisms frequently contain still more water. In that water, certain salts are dissolved, and they happen to be the very salts of sea water. So that's a good way to begin: with dilute sea water. In that medium of dilute sea water, one finds a special kind of molecule so closely associated with life that we call those special molecules *organic*. They are the molecules found almost exclusively as parts of, or as products of, living organisms, and they are made primarily of those four elements I spoke of last time: carbon, nitrogen, oxygen, and hydrogen.

That faces us, however, with an immediate dilemma: one can hardly conceive of making living organisms without a good supply of organic molecules, but, as I have just said, we hardly *know* organic molecules except as products of living organisms. It's the old question of which came first, the chicken or the egg. There's a curious confusion about this dilemma: After scientists had realized that these organic molecules so profoundly associated with living organisms are also *made* by living organisms, a German chemist named Friedrich Wöhler synthesized the first organic molecule, urea, in 1828. Ever since then, chemists have been in the habit of saying, "Well, we thought up to that point that it took living organisms to make organic molecules, but Wöhler showed that life was not essential at all, that organic molecules could be made synthetically." Yet really that's nonsense. It's as though one were saying that Wöhler was not alive, but of course he was. Organic chemists are just a curious class of organisms that make organic molecules externally as well as internally; the rest of us are content with making them internally. Virtually all organic molecules are made by living organisms, including organic chemists.

So the problem is how the Earth got a supply of organic molecules before it had any living organisms, and it turns out that rather ordinary geological processes, processes that happen on the Earth, in its atmosphere and in its waters, can generate organic molecules. In the early 1950s, a young student named Stanley Miller did an experiment that immediately became famous — at one stroke it changed our whole concept of how difficult or how easy it might be to get organic molecules by rather simple, inorganic, non-living processes. Miller circulated a mixture of four gases through a closed glass vessel: methane (CH_4), the simplest of carbon compounds; hydrogen gas (H_2); water (H_2O); and ammonia (NH_3). He passed this mixture of gases through an electric spark for a week. At the end of that week he found that in one of his experiments more than half of the carbon that had gone into the experiment in the simple form of methane had been converted to a great variety of organic molecules, including twenty-five different amino acids — some of them the very kinds that engage in making proteins, other organic acids, and a wide variety of other molecules.

Why did Miller choose those four gases for that experiment? For the good reason that those four gases are believed to have been most prominent in the primitive atmosphere of the Earth. The reason they were so common, as I pointed out in the previous chapter, is that when a new sun and its planets are formed, they're formed largely of hydrogen. Our present sun is still composed 99 percent of hydrogen and helium. All the other elements together make up only one percent of its substance.

So one expects — as Harold Urey, Stanley Miller's teacher, argued — that generally in the early atmosphere of planets there will be a lot of hydrogen. As long as that hydrogen remains, the other atoms tend to be combined with hydrogen: the oxygen as water, H_2O; the nitrogen as ammonia, NH_3; the carbon as methane, CH_4. Those four gases mainly form the primitive atmosphere of almost any planet and are believed to have formed the first atmosphere of our planet.

An atmosphere is held by gravity, but the lighter gases keep leaking away and being blown away by the sun's radiation, the so-called solar wind. The lightest of the gases is hydrogen itself, so it is lost first. As the hydrogen escapes, the atmosphere begins to change. It is increasingly made up of heavier gases, and those elements we spoke of become combined, sometimes with themselves, sometimes with other elements, but no longer with hydrogen, so that we find the nitrogen, as at present, mainly as nitrogen gas, N_2; the carbon as carbon monoxide, CO, and carbon dioxide, CO_2. That change occurred in the atmosphere of our planet several billion years ago. By now experiments like that of Stanley Miller have been conducted using combinations of all those other gases — twenty different combinations in all. Always one gets exactly the same result.

That's where the unit organic molecules came from. They were mainly bred in the upper atmosphere through reactions sparked by lightning discharges and ultraviolet radiation from the sun. Over the ages they formed slowly, yet continuously, and were leached out of the atmosphere by the waters of the Earth, by the oceans, so that the oceans accumulated larger and larger concentrations of organic molecules of greater and greater variety. Those molecules, dissolved in the waters of the seas, collided with one another, and in colliding they reacted with one another, so the variety kept increasing. Some of them began to join together to make bigger molecules, sometimes adsorbed on the surfaces of rocks and clays. So one got not only new, but larger and larger organic molecules. Those larger organic molecules moved more slowly and were sticky, and sometimes in their collisions they clung together, forming aggregates of molecules. The Russian biochemist Alexander Oparin has suggested that natural selection already began at the level of these aggregates: by virtue of their composition or organization, some were more efficient than others at incorporating further organic molecules, so those aggregates grew at the expense of the others.

It appears that sometime, somewhere, or several times in several places, such a complex aggregate of organic molecules in sea water reached the kind of organization that we would recognize, had we been there, as life. That seems to have happened on Earth about three billion years ago. Having achieved such a foothold on life, the first organisms were launched upon the grand enterprise of evolution.

With that first appearance of life, one encounters the next problem: how those first primitive organisms could have lived. For life is not a static phenomenon. Any living organism is the seat of a continuous flow of matter and energy, which constitutes its life. Once that flow stops, the organism is at least dormant, or more likely dead. Now we come to a very important consideration. Arising as those organisms did in a soup of organic matter, with no oxygen in the atmosphere, they could have lived in only one possible way: that is, by fermentation, or what Louis Pasteur called "life without air," meaning life without oxygen. The problem of those first organisms was how to get hold of the energy they needed in order to live without the presence of oxygen, and they could do that only by fermentation. In fermentation, an organism takes an organic molecule — sugar — and, adding nothing and subtracting nothing, rearranges that molecule's atoms so as to squeeze out a little energy on which it can live.

Sugar is a rather small organic molecule; it has the formula $C_6H_{12}O_6$: six atoms of carbon, twelve of hydrogen, six of oxygen. The fermentation most familiar to us is performed by yeast, and what the yeast does to the sugar is break it into two molecules of ethyl alcohol and two of carbon dioxide — don't forget that carbon dioxide; we're going to need it in a few moments. The yeast does not do that to get rid of sugar or to make alcohol, but to get a little energy. From 180 grams of sugar degraded in this way, it derives about twenty thousand calories of useful energy.

The problem with fermentation is that in the long run it is a losing game. Those first organisms, having arisen as they did in a soup of organic molecules, were now turning upon those organic molecules and beginning to devour them. Clearly this process must eventually come to an end, just as we, who some time ago began to devour the accumulation of past ages of fossil fuels, are beginning to come to the end of that game. Those organisms consuming the past ages of accumulations of organic molecules eventually would have had to run out of them. With that, life would have come to an end, and the whole process would need to start again at the beginning. Fortunately, before that happened, by using the carbon dioxide that had been a by-product of fermentation, thrown in increasing amounts into the atmosphere, and using the water that was all around them, organisms invented a new process: *photosynthesis*. In photosynthesis, an organism whips together six molecules of carbon dioxide with six molecules of water and, using the energy of sunlight, makes of them a molecule of sugar and a by-product of the highest interest: six molecules of oxygen gas, O_2. We'll come back to that oxygen in a moment.

The development of photosynthesis was probably the greatest single event in the history of life on this planet. Through it, organisms became independent of their history. Up until then, they had been completely dependent on ages of accumulation of organic molecules. Now, using sunlight, they could make their own organic molecules. With that, their future on Earth was assured.

Then there was the by-product, oxygen. As organisms began to pour oxygen into the atmosphere through photosynthesis, they could begin to use that oxygen to burn the new organic molecules they were making, the sugar. So those organisms invented another way to live, which we call respiration. Respiration is a form of simple combustion. It is the burning of organic molecules with oxygen, just as one might burn sugar by lighting it with a match. In just that way, but at low temperatures, the organism, instead of generating light and heat, generates the energy that it needs to live. From this process of respiration, the organism gets about twenty times as much useful energy as it could have gotten from that same amount of sugar by fermenting it, so respiration was an enormous release for organisms. Fermentation, the first way in which living cells learned to live, still remains their basic form of metabolism. Few, if any, living cells have ever forgotten how to ferment. But most of them by now, having access to oxygen, add respiration to fermentation. No organisms that have gone on living entirely by fermentation have ever amounted to much. It's a subsistence way of life, a poor way to live. It uses a lot of sugar and yields very little energy. Respiration generates a great surplus of energy, and that surplus was invested in the enterprise of evolution.

Oxygen, the by-product of photosynthesis, did another very important thing for organisms. Sunlight includes shortwave, hard, ultraviolet radiation that is incompatible with life. It kills all living organisms and even destroys the essential structures of proteins and nucleic acids. As long as that hard, ultraviolet light reached the Earth, organisms had to stay underwater. But some of the oxygen that was thrown into the atmosphere as a by-product of photosynthesis was again activated by sunlight and turned into ozone. (Oxygen gas is O_2, ozone is O_3.) That ozone, high up in our atmosphere, is what blocks the hard, ultraviolet light from reaching the Earth. There isn't much of it — all the ozone in our atmosphere, if compressed to one atmosphere pressure and brought to zero degrees centigrade, would form a layer only three millimetres thick. Yet that's enough to keep the shortwave, ultraviolet light from reaching the Earth. It was only after we got that thin layer of ozone that organisms, at last, could come out from underwater and populate the land and the air.

So that's the way life was launched upon the Earth. By now it has multiplied and diversified to fill every niche, every crevice where it can find some proper substance on which to live and some source of energy. And in the course of populating the Earth, life has utterly transformed it in very important ways.

Ordinarily, even biologists are in the habit of thinking of the physical environment as the thing given, as fixed, and life as having somehow to fit itself into that fixed environment. The environment is thought to play the tune to which life must adapt, must either dance or die. But we realize now that it hasn't been that way at all. Some of the most important aspects of our physical environment on the Earth are the work of living organisms. I have just told you of one of them: all the oxygen in our atmosphere was put there by plants in the process of photosynthesis, goes into plant and animal respiration, and so is completely renewed every two thousand years — and two thousand years is just a day in geological terms. With carbon dioxide it's even stranger: all the carbon dioxide in our atmosphere, and all of it that is dissolved in all the waters of the Earth, goes into photosynthesis and comes out in plant and animal respiration, and so is completely renewed about every three hundred years. The strangest thing of all is water, because all the waters of the Earth go into photosynthesis and come out of plant or animal respiration, and so are completely turned over, completely renewed, every two million years — and two million years isn't long in geological terms. So we see that the whole surface of our Earth and its atmosphere are involved in the metabolism of living things and have been utterly transformed by that process.

One doesn't give life its due. We think of life as a newcomer: a newcomer to the universe, to this planet, something tender and evanescent, something of the moment as compared with those age-less stars. That isn't so. The bigger a star, the more quickly it burns up its hydrogen, and so the shorter its life, the shorter its time of maturity on the main sequence. Our sun became mature, entered the main sequence, about six billion years ago, and has another six billion years to run. But a star of twice the mass of the sun would have a total life on the main sequence of only about two billion years. That's a shorter time than life has existed on the Earth. Life started on this planet some three bil-lion years ago, and many a star has been born and died since then. A star twenty-five times the mass of our sun — and there are such stars — would have a lifetime of only about three million years. That's very little; we've had man-like creatures walking the Earth that long, and some stars have been born and died since creatures like men first began to walk the Earth. Perhaps I should add that a star half the mass of the sun would have a lifetime on the main sequence of about fifty billion years. Fifty billion years: that would be a time full of promise, instead of the paltry ten to twelve billion years that have been allotted to us with our star.

Life is a great thing, whenever and wherever it appears. Vast in its effects, rivalling the stars, it has indeed everything, every attribute but mass, and if one has no overweening respect for mass, life has dignity enough. The entire point of this story has been to bring this improbable phenomenon, the origin of life, into the order of nature, to make of it one of those repetitive events, something that will happen, given enough time, wherever it can, wherever the conditions permit. What are those conditions? Well, I think it's quite clear that for life to arise it takes a planet something like the Earth, of about that size, that temperature, and receiving about that amount and quality of radiation from its sun.

How many such planets are there? Well, this is a vast universe. Even the parts of it are almost unimaginably vast. For example, our own home galaxy, the Milky Way, has a cozy nearby sound to it; yet it takes light travelling at 186,000 miles a second something like one hundred thousand years to cross the Milky Way from edge to edge. Many of us are beginning to feel crowded on this planet with a world population of about 3.5 billion people, but that home galaxy of ours, the Milky Way, contains about one hundred billion stars like our sun. The lowest estimate I have ever seen of how many of those stars are likely to have planets that could support life is one percent. Take one percent of one hundred billion, and that's a billion such planets in just the Milky Way. And since our galaxy is just an ordinary, run-of-the-mill galaxy, and there are something like a billion such galaxies within a radius of a billion light-years of the Earth — well within the range of our most powerful telescopes — a very conservative estimate would say there are perhaps a billion billion such planets capable of supporting life within the already observed universe. Life may not exist in all those places, because in some of them enough time hasn't elapsed, but it must exist in many of them. And that is an important realization: we live in an inhabited universe, one in which life exists in very many places.

You start a universe such as this one with hydrogen, or with neutrons, or with radiation, and given enough time, life will arise. It's part of the order of nature and, having arisen, life will begin to evolve, for evolution, too, is part of the order of nature. Life takes a high place in that order. So far as we know, it represents the most complex state of organization that matter achieves in our universe. And here in our corner of the universe, through a long journey, an immense evolution, stuff cooked in the deep interiors of previous generations of dying stars and gathered up over endless ages of time from all the corners of the universe came finally to form a creature that has begun to contemplate itself and to ask such questions as those we are asking here.

That brings us, at last, to man — the animal that knows, the animal that creates, the science-making and art-making animal. I shall speak of him in the next chapter.

THREE

THE ORIGINS OF MAN

WE ARE GOING ON a long journey, and the point
is to know what that journey is and to ask what it
means. The universe appears to us in a hierarchy
of states of organization. The matter is always the
same; it is the organization that's different. As the
organization grows more complex, new properties
or attributes appear, and those are what interest us.
That hierarchy of states of organization begins with
the elementary particles; then the atoms; then, of
them, the molecules; then living organisms; then
plants and animals; and then societies. In our historical universe, this is also the order of succes-
sion in which these things appear through the ages.

Ten to twenty billion years ago, our galaxy was
formed. It is a galaxy like many others; already

perhaps one billion such galaxies are visible to the most powerful telescopes. About six billion years ago, our sun was formed. It is one among many such stars; something like one hundred billion exist in our galaxy alone. About 4.5 billion years ago, the Earth took its present form as a planet like many others. About three billion years ago, life came to the Earth, as it has to many other places in the universe — perhaps to as many as a billion such planets in our own galaxy. Then, two million years ago or so, man began to walk the Earth, and that's where we've come to in our story: the coming of man — but the coming of man out of this larger history, as part of this history, as a kind of culmination of it.

About four hundred years ago, a collection of molecules organized as William Shakespeare wrote *Hamlet*, and about a century later, another such collection of molecules organized as Isaac Newton wrote the *Principia*. Then, about a century and a half ago, another such collection of molecules organized as Ludwig van Beethoven wrote the Ninth Symphony. I do not say this kind of thing to disparage man; I say it to exalt the molecule. When one sees a collection of molecules writing *Hamlet*, that shows what molecules, properly organized — that is, organized as William Shakespeare — can do. One has learned a little more chemistry.

It's a curious thing that some persons think one can take a nobler view of these phenomena. It is commonly conceded that if a man has just eaten and is digesting, we can relegate that process to his biology and chemistry. If he's moving, well, that's some more biochemistry with some physics. If he is excreting wastes, well, the chemists and physiologists can have it. But when that man writes a sonnet, the impulse is to say, "Ah! That's something else. That's coming out of some higher thing. That is not his body, but his spirit at work." That spoils the whole story, though, because it makes that man something like a radio, just a receiver for a message from outside, something transmitting what is not his own. But it is his own: the sonnets, just like the enzymes in his digestive system, just like the working of his muscles, are coming out of him. It has taken a long, long time, but start a universe like ours with hydrogen and wait long enough, and eventually there are places where molecules have formed and have made living organisms; the living organisms have evolved, and then one comes to a time when one gets such a thing as a man, and he writes sonnets. It's one world.

In an increasingly technological world, one tends to ascribe everything that exhibits design to technology. One sees a car and asks, "Who made it?" And the answer comes back, "General Motors" or "Chrysler" or whatnot. One sees a refrigerator and asks, "Who made it?" and the answer comes back, "Frigidaire" or "General Electric" or whatnot. So one sees a man and asks then, "Who made him?" and the answer given sometimes is "God." But, in fact, our world faces us with two quite different kinds of processes for achieving design: *technological* and *biological* — or, as I think better to call it, *organic*. These two processes are entirely different from each other. Indeed, one is almost the reverse of the other and, particularly now in these confused times, it is of the highest importance to sort them out and recognize which of them one is dealing with in any given instance.

Technological design begins with specifications. One writes the specifications and then tries to achieve them in any way one can. But *organic* design has no specifications. It is an entirely different kind of process, a process that was characterized for us a little over a century ago by Charles Darwin and called by him "natural selection." It has three components: variation, inheritance, and what Darwin called "the struggle for existence." Organisms are subject to variations, mutations, and differences of every kind, some advantageous and some disadvantageous, and some of those variations are passed on to other organisms. The process of inheritance can be biological, genetic, or, in creatures with societies or with the habit of retaining a long association with their young, it can involve the transmission of learned behaviour, cultural inheritance.

I should add that there are few tasks as difficult as determining in any given instance which of those forms of inheritance one is dealing with, because they mimic each other so closely. For example, some researchers have alleged that black people statistically exhibit lower intelligence than whites, and that this is owing to genetic causes. That lower intelligence is certainly not in my experience — I am a teacher and I have increasing numbers of black students, and they are superb — but we are told that certain tests show statistically a lower intelligence among blacks. The problem here is what is being tested and then whether it really is genetic. Many of us are inclined to think that whatever it is being tested — and that is by no means clear — is social in its causes rather than genetic; that it is a matter not of nature, but of nurture.

So, this matter of natural section has three components: variation; the mechanism of inheritance, whether physical or cultural; and the third component, a competitive, selective factor, "the struggle for existence." In the long run, every population of living things, plant or animal, begins to exceed its resources. At that point, those living things are thrown into competition with one another for the resources they need, and in that struggle, in that competition, the things that work a little better survive and the things that work a little less well perish. This is the process Darwin called "the survival of the fittest." In this process, there are no specifications, no foreordained design, and hence no need for a designer. It is a process of editing rather than authorship, and those who like such phrases could more accurately speak of the Great Editor of our being rather than the Great Author.

At first glance this process of organic design seems slow, inefficient, and wasteful compared with what we know of technological design, but we should think well of it because it has made all the organisms we know, including men. And even the simplest of living organisms is ever so much more flexible in its adaptations to changed conditions — and, to cite a more modern value, is ever so much

more highly miniaturized — than the most complicated and sophisticated of technological devices.

Sometimes I am asked whether human evolution still continues and, if so, in what direction it
is going. I am sure we are still evolving — as long
as we go on, we'll continue evolving — yet it would
take a braver man than I to try to say where it's
going. I would venture only a few comments, and
those rather trivial: We only lately lost our tails,
and indeed our skeletons still retain a vestige of a
tail, our coccyx. We also seem to be losing our toes;
they once were good for grasping and very useful,
but already they have little more than ornamental
value. If we manage badly in the years to come, so
badly that we have to take again to the trees, those
losses will prove a serious embarrassment.

It's curious that some persons suppose that there
is less dignity in this evolutionary view of the origins of man than if he were a technological product,
albeit one made by God. To me, the evolutionary
view is by far the nobler view. It has taken so long,
so much has gone into it, and it is bound up with the
whole history of the universe. It is the most moving, the most awesome, the noblest story I know.

Darwin recognized that in natural selection he
had laid hold of a universal principle, and in his

second book on evolution, *The Descent of Man*, he didn't hesitate to apply it to all kinds of human and social phenomena — to human ethics, aesthetics, the wars of nations, the competition of cultures. I should like to give an example of natural selection working in the social sphere: all of you know that Britain has no written constitution — the English never wrote a constitution. In what became the United States, those people in a new land, facing a wilderness, in their insecurity, wrote a Constitution, and it has made a lot of trouble since. But underlying that Constitution, underlying what one recognizes as the fundamental law of the United States and the fundamental laws of Canada, Britain, and many other countries, is the imposing structure of the English Common Law. That Common Law is a genuinely organic structure, a system of tradition and precedent built up over centuries of experience, of trying everything, all kinds of procedures, retaining what worked a little better and discarding what worked somewhat less well. That is the biological way of doing things, as contrasted with technological systems of law drawn to specification, such as the Roman Law, the German Law which is based on it, the Code Napoléon, and other such systems.

The closest realization of natural selection working in the political sphere would be democracy, and the indispensable element in democracy is its open-endedness. When it is working properly, it is endlessly experimental — trying everything, keeping what has been found to work somewhat better, and discarding what has been found to work less well. Our problem is to achieve that degree of democracy; but if we could, I would vastly prefer it to any such technological system as those planned societies, the communist or fascist utopias, that so captivated my generation.

I should like to say a further thing, which is that uniformity offers evolution nothing. Uniformity is the death of evolution, of all further development. Evolution absolutely depends on that ceaseless outpouring of variations, of individual differences, and this is the best argument I know for tolerance — but tolerance isn't enough. Tolerance is only a negative kind of response. When one realizes the importance of variations, of differences, and that all further development depends upon them, then one realizes that one must do more than merely tolerate them. They are precious, and one needs to foster them, to cultivate them. They are intrinsic to the operation of evolution and of democracy. They are the source of all genuine social and political progress.

I have been told, since I began studying science, that science is ethically neutral, that it has nothing to do with good and evil, with right and wrong — that it's concerned with what is, not with what should be. I think that's a superficial view of both science and ethics. Our ethical principles come to us mainly in the form of axioms stated as categorical imperatives. They have the force of Martin Luther's "Here I stand; I cannot do otherwise." They seem wholly arbitrary, but I think on examination they prove not to be all that arbitrary. Every culture survives or fails on the basis of its fundamental affirmations, and no society could long survive a choice of faulty affirmations. Take those Ten Commandments in our own culture: the first four are tenets of Judeo-Christian theology, and in any other theology would be different, but not the last six. Those last six are the rules by which a society even remotely like ours lives. I have sometimes tried the game of inverting them, of saying them in reverse. That inverted series of commandments would read, "Kill, fornicate, steal, lie, try what you can to lay hold of your neighbour's wife and his belongings," and one that has particular relevance for our time, "Despise your father and mother."

Well, just look at that inverted series of commandments, and you realize that no society in any way like ours could possibly survive if those were the rules. No, they have to be as they are, and that is why, if you venture into an entirely different time and an entirely different culture — if you go to Buddhism, Hinduism, Zoroastrianism, or what have you — there is no quarrel. For they have societies, too, and hence they too have to obey those last six commandments. Therein lies the real sanction of those commandments, rather than in a supernatural edict.

It's curious that anyone would think this organic view of man and his ways and institutions demeans him. It's curious that anyone would consider it nobler to think of man as a technological product, albeit made by God. The organic or evolutionary view seems to me to accord man the highest dignity: made of the stuff of stars, gathered up over vast ages of time from the remotest corners of the universe. All the history of the universe has gone into him and ended in making this thing, a man.

Without such a thing as a man, the universe might still *be*, but not be *known*, and that is a poor thing. In man and his like, matter has achieved a state of contemplating itself, of beginning to understand itself. You might have heard it said that a hen is just an egg's way of making another egg. Well, in exactly this sense, a man is the atom's way of knowing about atoms. A man, if you wish, is the star's way of knowing about stars — a great thing. I think we've had enough of humble man. I think he has good reason to be proud, not of himself, but of his kind; not in competition with his fellows, but of his fellows. Proud of men, he does not have to seek a spurious dignity; his place in nature gives him dignity enough. He needs no mystical revelation to assure him that he is one with the universe, for man *is* one with the universe, made of its stuff; its history, his history, ancient as some stars are ancient. In his contemplation, the universe contemplates itself; in his knowing, the universe knows itself; in his creating, something new is added to the universe. If man were to die, a little of the universe would die. Indeed, to borrow another form of speech, if God were dead, that would be because man had died.

FOUR

THE ORIGINS OF DEATH

I HAVE SPOKEN TO you of the origin of life and now I should like to speak of the origin of death and its meaning. Death is one of the great events of life, one of the great events of passage. And I can tell you beforehand that this will be the happiest of these talks.

Stars die, men die, but not all living creatures die. An amoeba, for example, need never die. If it does die, that is because of some untoward circumstance, some violence. Ordinarily, it just divides and becomes two new amoebas. That's the way it is with all such single-celled living creatures — they just go on dividing endlessly, making new copies of themselves.

Sometimes these creatures do things a little

differently. A well-known biologist of the generation before mine named Lorande Woodruff stud-
ied a variation on this process practised by some
single-celled creatures called *Paramecium aurelia*, a type of protozoan. Every now and then two
paramecia come together side by side, the cuticle
between them breaks down, and they exchange
some genetic material, then break apart again and
go right on dividing as they had done before. This
process is called "conjugation," and Woodruff was
studying whether paramecia could just go on living
by simple division without conjugating occasionally. So what he did was each morning he'd come
into the laboratory, and he'd find usually two paramecia where he'd left the one swimming around
in a bowl all by itself the night before. Then he'd
carefully separate them, so that each paramecium
was alone and there was no chance to conjugate.
After a while he began to write a series of scientific papers. The first one was called something like
"Five Hundred Generations of *Paramecium aurelia*
without Conjugation." Then, after a while, there
was another paper called something like "Twelve
Hundred Generations of *Paramecium aurelia* without Conjugation." And this went on year after year
until finally there was a culminating paper called

"Eleven Thousand Generations of *Paramecium aurelia* without Conjugation." In this way Professor Woodruff lived a happy and useful life and convinced us all that paramecia could go on and on just reproducing by simple division, without any conjugation.

However, in the course of watching these paramecia closely, Woodruff discovered a wrinkle in this process of reproduction. Now and then a paramecium, all by itself, would let one of its nuclei degenerate and make itself a new nucleus of the other one. You see, every paramecium has a little nucleus and a big nucleus, a so-called micro-nucleus and a macro-nucleus. The macro-nucleus is its working nucleus; the micro-nucleus is just a kind of reserve of genetic material. What he discovered was that if you keep paramecia all by themselves, after a while in sheer desperation one of them will perform a kind of "do-it-yourself" conjugation. It will let its macro-nucleus disintegrate, make itself a new nucleus out of the micro-nucleus, and then it's a brand-new paramecium. It's quite a trick.

There are all sorts of these variations, practised not only by this organism, but by many similar organisms — and yet, no dying. The curious thing is if one takes a big jump from such single-celled organisms to multi-celled organisms, for example to such a thing as a sea anemone — a great big thing compared to a single-celled organism. Sea anemones reproduce as a usual thing by just splitting down the middle, and then one has two sea anemones, as good as new. That kind of process is rather a habit at this level of organism. A close relative of the anemone, the hydra, reproduces by producing little buds off the side of the parent organism. The bud grows for a while, then it splits off and begins to live by itself, and one has a new hydra, as good as ever.

One can take another big jump and one is in the flatworms. Flatworms are a lower order of worm and yet they, unlike the hydras and sea anemones, are more like us in a way. They have a right side and a left side; they are bilaterally symmetrical, as one says. They have a head end and a tail end. They have a nervous system somewhat concentrated toward the head end, and they are reasonably complicated organisms. Yet a very common flatworm called *Planaria* reproduces as a regular thing by just pinching in at the waist and eventually breaking in two, and then each half regenerates whatever it lacks. The head end regenerates a new tail, the tail end regenerates a new head, and you've got two flatworms where you started with one.

There was a Dutch scientist named Stoppenbrink who some years ago started the curious experiment of starving these flatworms. He had a culture of *Planaria* and just kept them without food. When one does that, these *Planaria* begin to live on their own substance, begin to feed on themselves, and they do so according to a perfectly definite program. The first thing they do is get rid of any sexual products they happen to have. Then they absorb their own digestive systems, which under these circumstances aren't doing them much good anyhow. Then they begin absorbing their muscles, and they go on with this process, getting smaller and smaller and smaller. There's only one kind of organ that they don't attack, and that's the central nervous system. So, as they get to be smaller and smaller, that central nervous system keeps its size, and they end up looking very intellectual, all head and no worm.

Reading this account, I was dying to know whether Professor Stoppenbrink came into his laboratory one day and found that his *Planaria* had just faded away, that there were no more *Planaria*. Unfortunately, just before that seemed about to happen, Stoppenbrink began feeding those *Planaria* again and they promptly got back everything they'd lost. Well, I was deeply disappointed, except that in doing that, Stoppenbrink learned something new: the *Planaria* that came back on feeding were brand-new *Planaria*. You can, in fact, keep *Planaria* alive indefinitely by just periodically starving them and then feeding them again. There's no doubt a moral to that story.

The highest I've ever been able to pursue this method of reproduction by breaking up in to fragments, by dividing, by fission, is into the real worms, the cousins of our own common earthworm. There is one of them with the beautiful name *Enchytraeus fragmentosus*, and it got that second name because the only way it reproduces is by fragmentation. At the right time, it breaks into a lot of pieces, and each piece gains back everything that it lacks by regeneration, and you have five or six new worms where you started with one.

However, long before that stage, living organisms have achieved a quite different way of reproducing: the sexual mode of reproduction. And it is in the closest, most intimate association with the sexual mode of reproduction that death comes upon the scene. I can speak of this association best in the terms that a great nineteenth-century biologist named August Weismann spoke of it. He explained that every organism that reproduces sexually begins its life as a single cell, a fertilized egg. Then this cell begins to divide repeatedly, and in its divisions it eventually makes the whole organism. As the divisions go on, the cells begin to differentiate into various kinds of cell, and over the course of this process some of those cells are segregated off as what Weismann spoke of as the "germ plasm," and eventually produce the mature germ cells, the eggs and sperm. Other cells, in the course of these divisions, are differentiated to produce the body, what Weismann spoke of as the "soma." All these cells are coming out of that same original single cell, the fertilized egg, by the simple process of repeated division, but we end up with these two categories: the germ plasm that produces the eggs and sperm, and the body, the soma.

Now, using these terms, Weismann stated two great basic principles: the first of them he spoke of as the "isolation of the germ plasm." The way we'd say this nowadays is that genetic information goes in only one direction, always from germ plasm to soma, never in the reverse direction from soma to germ plasm. That's why there can be no inheritance of acquired characters. An acquired character is a difference in the soma, a change in the body, and there is no way that change can be transmitted to the germ plasm and hence inherited.

The other great principle was what Weismann spoke of as "the potential immortality of the germ plasm." You see, germ plasm — eggs and sperm — makes some more germ plasm and also makes a body. The line of germ plasm is potentially immortal: germ plasm goes on making more germ plasm endlessly, by simple division. But it also makes bodies, and the point of the body is to carry that germ plasm: to feed it, to protect it, to warm it (in warm-blooded creatures) and eventually to mingle it with the germ products, eggs or sperm, of the opposite sex, to start the next generation. Then the body has done its work and can be discarded, and that is what death is. Death is the discarding of the body after it has performed its function in the act of reproduction.

The thought that life is through with the body when reproduction has been accomplished is repugnant to us as men, and I shall have more to say of that a little later. But repugnant or not, let me say now: this would be no surprise to a salmon. In salmon, trout, eels, and many such creatures, the preparation for reproduction is simultaneously a preparation to die. The two things are happening

together, and reproduction is all too clearly the last
act of life. The life cycle of a salmon involves being
spawned in fresh water — always in fresh water —
and then, as it grows, it goes through a first metamorphosis: it loses its bright colours, turns silvery,
and usually goes out to sea. At sexual maturity, it
comes back into fresh water again, it migrates up
the rivers, going up to its spawning grounds. Pre-
paratory to starting that migration back upstream,
the salmon goes through a second metamorphosis:
it changes its colour again, the flesh pinks up, a lot
of anatomical changes happen, and one of the most
significant changes is that the whole digestive system degenerates and becomes unable to function.
That salmon is through with eating. It has only
one piece of business before it, and that is to get
itself upstream to the spawning grounds, shed its
sex products, and then die. It's no longer fit to go
on living. If you were at that point to cram food
down its gullet, it would have no apparatus with
which to use it.

That's the way it is with many such creatures.
Take, for example, what we call freshwater eels. All
the eels of the Atlantic, both the American and European, grow up in fresh water for a period of five to
fifteen years, then metamorphose at sexual maturity

and start off on an enormous migration which brings them all together in the deepest and saltiest part of the Atlantic Ocean, the Sargasso Sea. There, at very great depths, they spawn and die. And once again, before beginning that journey, all those eels go through a whole series of anatomical changes, get really stripped for action, stripped for that tremendous journey — and part of the stripping, once again, involves a complete degeneration of the digestive system. They'll never eat again. They don't eat in the whole course of that journey. There's only one thing before them, and that is to get to the Sargasso Sea, spawn, and die. Then the baby eels have to make their way back home alone, and how they get there is quite a wonderful and mysterious story, because the American eels and the European eels are different species but, so far as I know, no baby eel has ever got so balled up as to come to the wrong continent. It takes the American eels about fifteen months to migrate from the Sargasso Sea before they finally approach our shores, metamorphose, and come up into fresh water. It takes the European eels about three years to get home, but they do.

In the great sexual migrations, the migrations that go with the reproductive processes, it's all too apparent that reproduction is the last act of life, and the preparation to reproduce is simultaneously the preparation to die. There are other great migrations in which death also enters, but actually in those instances death is the objective. In the case of the reproductive migrations, death occurs because the organisms, the bodies, have fulfilled their function. In these other migrations I'm about to speak of, the whole point of the migration is to kill off the organisms.

These are the well-known hunger migrations. Under conditions of shortage of food, animals begin to move; they appear in all kinds of strange places where ordinarily they've never been seen, such as towns and other inhabited places. The most famous of those migrations is that of the lemmings. Lemmings are rather large rodents that ordinarily live high up on the mountainsides of Norway. Every few years, there is a great lemming migration, and you've all heard the mythology about it. Great hordes of lemmings, perhaps hundreds of thousands of them, come pouring down off the mountainsides of Norway, and enter the cities, stop all the traffic, and they're on their way to the sea. When they get to the shore, they plunge into the ocean and swim out in an act of mass suicide, never to be seen again.

It's an impressive story, but as I understand it, the reality isn't quite that dramatic. What is happening is that in a so-called "lemming year," a year of lemming migration, the population of lemmings in a particular area has outgrown its resources. There isn't enough food to go around, and there's a built-in, automatic response to lack of food, to genuine deprivation of food, real hunger, that one finds in all living organisms, from single-celled protozoa right up to man: it is that hunger drives the creature to run, to move, to wander, and the wandering doesn't have any particular direction. It's just wandering, and that's what's happening to the lemmings. Such is the shape of Norway that a lot of lemmings pretty soon hit the seashore that is on one side of the mountains and plunge into the ocean and drown. But on the other side they hit the plains of Lapland and wander off across those plains to die. There is this great mass of lemmings drifting away from their centres of population and dying, and that's the whole point of the process: there is a shortage of food; there are hungry lemmings; the hungriest lemmings begin to wander; they wander away and die; and when enough lemmings have wandered away, there is now enough food for those that are left, and automatically the migration ceases.

It's rather curious that we, as men, think of this as a kind of aberration. There's something that we reject in it; we feel that the suicide of the lemmings is highly abnormal behaviour. This is a rather curious attitude because, where hungry lemmings go off to die, hungry men go off to kill. We seem to think our way is more natural, but there is a lot to be said for the lemming way of doing things. For one thing, it ends in a lot less dying. The hungriest lemmings go off to die, but as I've just said, as soon as the population has sufficiently thinned out, the migration, and hence the dying, stops automatically, so there's a lot less dying. A second thing: there's no destruction; the lemmings that survive have their homes intact, just as before. And a third: a selection process is at work, because it's the hungriest lemmings that go off to die; the ones that are doing better stay home. In the human way of doing things, we pick the flower of our manhood to go off to kill and die. Biologically, the lemming way of doing things is ever so much more effective.

Now I would like to speak of man in this context. First, let me deal with that repugnant thought that once reproduction has been accomplished, life is through with the body. Most of us don't like that thought, and indeed it doesn't apply to man — and since whenever a scientist says something uncomfortable, he starts explaining that it isn't so for man and tells some sort of Pollyanna story to cover it up and cheer up his listeners, let me speak instead of bees.

Bees have a society just as man does, and the very heart of that society, the thing that makes it go, are the so-called worker bees. They're sexless females: they'll never reproduce; they have no part whatever in the reproductive function. The only reproductive female is the queen; one queen per hive. But all the other business of the hive is done by those worker bees: they forage for food, they make the cells, they feed the grubs, they feed the grubs that are to become queens special food, they clean the hive, they air the hive, they do everything — and yet they're sexless. That's sort of the way it is with men, too: if there is a society, some of its components can serve the society, and the question of reproduction becomes irrelevant. Bach had a lot of children; Beethoven, so far as we know, had none. Who cares? It's utterly irrelevant. These aren't the things that we've come to Bach and Beethoven for. Newton had no children. Who cares? Utterly irrelevant. Those persons served our society, and served it very well. We're proud to be men because they once lived and did the things they did. That's the way it is with us — the question of reproduction can be quite irrelevant. One can serve society in many other ways.

Now, all through his history, man has pursued a dream of immortality. I'm speaking of immortality of the flesh, of the body, not of the spirit, whatever that might mean. So we have these age-old searches for the Fountain of Youth, the Philosopher's Stone, you've heard of them all. The curious thing about this is that if one achieved everything one could hope for in the way of immortality, it would make our condition so little different from how it is now as to be wholly uninteresting. I haven't the time in these talks to go into this matter in detail, but it's been written about very beautifully by Peter Medawar in a book called *The Uniqueness of the Individual*. In the first two chapters, Medawar lays out this whole situation: suppose one had everything one could hope for in the way of bodily, fleshly immortality — one might, for example, permit man to reach something like the maturity of a twenty-year-old and then never grow any older, and one might also give him the boon of no natural death. At this point Medawar becomes very interesting indeed, because all of us talk of natural death as though it were a real phenomenon. But Medawar writes that he asked all his physician acquaintances if they had ever witnessed someone dying a natural death — that is, dying of old age — and all of them loudly disclaimed it, said they had never seen this happen. There's always something else that supervenes: a last attack of pneumonia or some organ failure. Natural death is something we talk about, but it is highly questionable if it exists.

Nevertheless, for these purposes, let us have perpetual youth, no natural death, and then, says Medawar, let's give man a bonus: perpetual fertility. Let him reach the fertility that a twenty-year-old possesses, and keep it forever. And where would that leave us? Well, we'd still be subject to all the ills that flesh is heir to. We'd still be subject to disease, being run over in the street, war, pestilence, famine, all the violence — the insurance actuaries would just have to stand back for a little while and watch, and pretty soon they'd tell you the new rates. Matters would, in fact, have changed very little if we went on living the lives that we know men live — so little that Medawar suggests finally that the period allotted to us now, the period within which we grow old and die, is very close to the period we'd have if we were, in fact, immortal.

I should like to say a last thing: the strange part of all this is that we already have immortality, but in the wrong place. We have it in the germ plasm. We want it in the soma, in the body. That body has come to mean everything to us. We've fallen in love with the soma, the thing that looks back at us from the mirror. That's the repository of that identity that one keeps chasing throughout life. It's come to mean everything to us, and as for the germ plasm — that potentially immortal germ plasm that might go on making bodies ten thousand years hence, a million years hence — many of us couldn't care less.

I, too, used to have that thought. I, too, used to think that our immortality was in the wrong place. But I don't think so any longer. You see, every creature now alive represents an unbroken continuity of life, since life first appeared on this planet some three billion years ago. That really is immortality, because if that line of life had ever broken, how could we be here? Three billion years, and all that time that germ plasm has been reproducing by simple division. It's been living the life of those single-celled creatures, the protozoa, and all the while making and discarding bodies. If a germ plasm wants to swim around the ocean, it makes itself a fish. If it wants to fly in the air, it makes itself a bird. If it wants to listen to lectures on the radio, it makes itself a man.

That germ plasm that we carry in us now has been going on this way for three billion years, and during that time it has done all those things. There was a time when our germ plasm, the germ plasm that is in you and me, was making a fish. There was a time when it was making amphibia, things like salamanders and frogs. There was a later time when it was making reptiles that climbed out upon the land. Now it is making men, and if we only have the good sense and restraint to leave it alone, heaven knows what it will make in the ages to come. I, too, used to think we had our immortality in the wrong place, but I don't think so any longer. I think it's in the right place. I think this is the only immortality worth having, and we have it.

FIVE

ANSWERS

WHENCE DO WE COME, what kind of thing are we, and what is it to become of us? Those are now the burning questions, because we've come to a time of great decision, not only for men but for much of life on this planet. We men, who for many millennia have lived in symbiosis with other animals and with plants to mutual advantage — the men got along, the plants got along, and the animals got along — have begun to destroy life, to devastate the earth and the seas, and to extinguish other living things.

I'll begin with a parable; not a biblical parable, a biological parable. Some two hundred million years ago, we were in the age of reptiles. Dinosaurs were the lords of the Earth. They were the biggest land

animals that have ever existed. They were well pro-
tected: scales, armour plate. And they were well
armed: horns, teeth, claws. They looked very good,
those dinosaurs. And back in the shadows, hiding
among the roots of the trees, were a new, tender,
defenceless, small group of animals. They were the
first mammals. They had not much to offer, but one
thing: they had rather large brains for their size. A
dinosaur has a very small brain for its size: its proportion of brains to brawn is very low. The mammals were doing better in that regard, and pretty
soon there were no more dinosaurs. The age of reptiles had given way to the age of mammals.

The mammals grew and flourished upon the
Earth and kept working away on that beautiful
brain, and about two million years ago they gave
rise to men. Man is a lovely animal. You take him,
preferably unclothed, standing on his two feet,
strong, versatile, tall, gentle, and he is altogether
a fine animal, something one could love. But put
him in a car, and you can hardly see the man. He
is making a roar and a stink through the streets.
It's hard to love a man in a car, and once again the
proportion of brains to brawn has sunk very low.
He's like a medium-sized dinosaur, and dangerous.
Cars kill more than fifty thousand Americans each

year, more than we've yet lost in the whole of the
Vietnam War. And notice that we say cars kill those
Americans, not men in cars, knowing full well that
the men are not altogether in control. That's our
problem: while we weren't watching, we've become
dinosaurs again, and cars are only the beginning
of it. There are trucks and trains and planes and
hydrogen bombs, and computers run all of it. The
proportion of brains to brawn has sunk very, very
low and is sinking rapidly.

The mammals brought something else to life on
the Earth, something infinitely precious: mammals
take care of their young. Dinosaurs laid their eggs
and left them, and that was the way it was with all
the earlier creatures — the fish, the amphibia, the
reptiles — but not mammals. Mammals carry their
young in their bodies for long months, and then,
having borne them, they nurse them. That's where
the world "mammal" comes from, the nursing. Having weaned them, they watch them play in the sun,
protect them, feed them, and teach them the ways
of life. And that's another problem: we're no longer
taking very good care of our young. We've intro-
duced them into a world that offers them very little
they want and threatens their very existence. We're
becoming dinosaurs again — but this time if there's

an extinction, it will be a do-it-yourself extinction and there's no other creature to take over.

I am a scientist, but I am also a teacher. I teach about 350 young men and women at Harvard, and over these last few years I have realized increasingly that they've been more and more upset, more and more troubled. It's been harder and harder for me to teach them and, it seems, harder and harder for them to learn. Of course I've had to ask myself what it is that is troubling them, and when I was asked by a group of concerned students and faculty at the Massachusetts Institute of Technology to speak on March 4, 1969, I'd been thinking of these things and asking myself this question and I'd finally come to feel that I knew the answer. Then I had a dreadful problem, because it was a dreadful answer. So did I dare say it? I decided to say it, and I called that talk "A Generation in Search of a Future." I think that's the heart of the problem: those students I teach, and my own children, can't be sure that they have a future. And what is most dreadful is that I, a parent and a teacher, am in no position to reassure them.

We've come to a strange pass in cosmic history and the history of life on this Earth. I've heard the silly question asked, "Why is the world five billion years old?" And the silly answer is, "Because it took that long to produce a creature who would find that out." And now we have to ask a further question: Having reached that point, having achieved that creature, can we go on much further?

What's the trouble? Well, it isn't one trouble; it's a host of troubles, all interrelated and all coming to a head together within the next fifteen to at most thirty years — much too rapidly to hope to deal with any of them gracefully. A host of troubles, all of them frightfully complicated and all interrelated, so that if you want to deal with one of them, you must deal with all. Three of these troubles are overwhelming in their magnitude: the matter of pollution of the environment, the matter of overpopulation, and the matter of nuclear warfare.

Let's start with pollution. Pollution isn't a local problem, though it has its local forms. It's not a national problem. It's a world problem. DDT by now reaches into every corner of the Earth. The fish in the Arctic waters contain it in quantity, as do the seals. Eskimos haven't been using DDT — why should they? — but they're getting it in their food. I'd like to give you another example that is very serious indeed. I explained in a previous lecture that all the oxygen in our atmosphere — all the oxygen gas, the O_2 upon which we depend to breathe — is the product of living organisms. It was put into the atmosphere by plants in the process of photosynthesis, and that's all that keeps it there now. It may surprise many of you to realize that the great bulk of that photosynthesis, anything from 70 percent to 90 percent of it, occurs not on the surface of the Earth and in the plants with which we're most familiar, but in the upper layers of the oceans, performed there by algae. Now we are being told by experts, by scientists who have devoted themselves to finding these things out, that the oil residues present in those same upper layers of the oceans are equal in bulk to all the algae. We don't know at this moment what those oils are ultimately going to do to the algae, but we're taking a dangerous chance.

The environment has been spoken of as a motherhood issue — you know, everybody's for motherhood, though as we'll see in a moment, one can get too much even of that. Every politician, almost anywhere in the world, is willing to hold eloquent speeches about the environment until one asks him what he plans to do about it. That's where things get complicated, because there is that biggest water polluter, the oil industry, and the biggest air polluter is the motor car, which brings us to the auto industry, and oil again and gasoline. Another very big polluter is the lumber industry. Another is the power industry. So, as soon as one tries to do something about pollution, one encounters these enormously powerful forces, very wealthy forces, exercising great political power, running lobbies in government and so on. The only proper solution for the pollution problem, the one we must bring about, is to block pollution at its source. And that's going to be very difficult. That would be the human solution, and we're ever so much more likely to be offered the dinosaur solution, which is to let all that pollution go on and to superimpose on it a new multi-billion-dollar government-supported industry of anti-pollution. In these days of conglomerates, they would be the same business: one division would pollute and the other division would clean up, and both of them would make a lot of money.

The way I'm talking now — and I'm not alone among scientists in talking this way — raises rejoinders in certain quarters. Sometimes one hears disparaging remarks about "the prophets of doom," and there is a general impression that it's all a little hysterical, all exaggerated. We're told that we've always had problems and these are just our present problems. I wish it were so. These things I'm talking about are not in my field of science. I've heard of them and become deeply interested and read as best I could what information is available on them and I've gone to the experts, and I want to say something to you about that: the deeper one plunges, the more knowledgeable one's informant, the worse it sounds. I'm afraid, much as I would like to believe that what I'm saying is exaggerated or hysterical, it's not so at all. It is all too true.

So let me say a word about the matter of nuclear war. There, again, we are told, "Heavens, we've always had weapons. This is just a new weapon. Get used to it. Think of the mean things they used to say about the bow and arrow." But this is different. One could win wars with previous weapons. We've won such wars, whatever that might mean. One can't win a nuclear war. Both sides lose. It's entirely self-destructive and self-defeating. When I held that speech in March 1969, the stockpiles of nuclear weapons, both in the United States and the Soviet Union, had reached an insane level. Already the combined stockpiles had reached a level equivalent to the explosive power of fifteen tons of TNT for every man, woman, and child on the Earth. A general in the Pentagon with a beautiful name, General

Starbird, was asked to prepare an estimate for the Department of Defense of what the result might be of what they called euphemistically a "full-scale nuclear exchange" between the United States and the Soviet Union. It's curious, the phrases they use — we're all for international exchanges, and this is just one kind. General Starbird estimated that such a full-scale nuclear exchange would kill about 120 million Americans within one or a few hours. But that's nothing to worry about — we could simultaneously kill about 140 million Russians.

Those stockpiles were already at that insane level, and we are now in the midst of the next escalation, which is confidently expected to multiply those stockpiles by a factor of approximately five times, on both sides, both the Soviet Union and the United States, by 1975. Where does that leave us? As Senator Richard Russell of Georgia said on the floor of the Senate, "If we have to get back to Adam and Eve, I want them to be Americans, and I want them to be on this continent and not in Europe." That was an American senator holding a patriotic speech. Confronted with this situation, I constantly ask myself, was this inevitable? Is it inevitable that, having laid hold of a new source of energy, one uses it in this way? I'd like to say something about that.

The biggest event in the past history of life on this planet was the development of photosynthesis. Living organisms had arisen in a kind of soup of organic molecules and, having arisen, turned upon and began to devour those organic molecules in order to live. Inevitably that game would have had to come to an end, and life would have come to an end. But before that happened, organisms developed photosynthesis. With the energy of sunlight, they began to be able to make their own organic molecules. That made it possible for life to go on indefinitely on the Earth.

A second such great event has occurred, and that was the discovery of means of gaining access to atomic energy, to nuclear energy. That sunlight, on which all the life on Earth runs, is made by exactly the same nuclear process as occurs in a hydrogen bomb, a so-called thermonuclear reaction. And we are just in the midst of developing the technology of getting something other than an explosion out of that, by running it as a controlled reaction, as a controlled source of energy. That would mean that we could make our own sunlight, so to speak. We'd make life on Earth independent at last, even of the sun, and that could be a great liberating force. But instead of that, nuclear energy has become the greatest threat not only to human life, but to much of the rest of life on Earth. And so I have been asking myself lately: Is that inevitable? Does it have to be that way?

That's brought me to think rather fondly of the ancient Chinese. You know those Chinese, that mad and altogether unpredictable people, as Americans are taught to think of them. Those crazy Chinese, who many centuries ago invented glass, invented paper, did the first printing with movable type, and along about the eighth century, as I recall, invented gunpowder. And what did they do with that gunpowder? They made firecrackers. That's the way I've heard it. And so I think, "No, it isn't inevitable that, given access to energy that could expand life, that could make it ever so much richer, one uses it instead to kill."

That's another of our problems. We have to get rid of those nuclear weapons, everywhere. They're threatening our existence and the existence of much of the rest of life on this planet. And please don't misunderstand me: I am not suggesting unilateral disarmament. No, I want bilateral disarmament, indeed, multilateral disarmament. We're living now in a balance of terror and all that matters is the balance. I want those stockpiles of nuclear weapons reduced, all in balance, to half, to a tenth, to one percent, and then I would like to get rid of them entirely.

That brings me to the third problem: the population problem. Biologists, who like to measure all things, asked long ago how they were to measure fitness, that fitness spoken of in Darwin's phrase, "survival of the fittest." They decided that the best measure they could find of fitness was reproductive success: those organisms and those lines of organisms they would consider most fit were the ones that produced the largest number of offspring that survived to sexual maturity and in turn reproduced. Now I suppose man is the first creature on Earth, animal or plant, that is threatened by its own reproductive success. The world population has reached about 3.5 billion, and unless something strange and unexpected happens, it will double by the end of the century. Long before that happens, we can expect famine on an unprecedented scale in many parts of the world. But to go on feeding people adequately is not the heart of the problem — to think it is would be to accept that the great human enterprise, which has so much to offer, is now to become a meaningless, futile, bankrupt effort to see how many people one can keep alive on the surface of the Earth. That would turn the whole human enterprise into an exercise in production rather than creation. It would be utterly bankrupt and, however well one managed it, it would still need to come to an end. The surface of the Earth is limited and its resources, however we exploit them and however well we manage them, will eventually be exceeded.

Even if we could support this growing population, that is not the central problem. Our central problem is not the *quantity of men*, but the *quality of human life*. From that point of view I think we are already overpopulated, and not just in places such as India, China, Pakistan, and Puerto Rico, but here in the developed countries of the western hemisphere and of Europe.

We have made other mistakes in this regard. All my life I have been told that the problem of overpopulation is essentially a problem of over-reproduction of the poor. I have been told that the difficulty is that the poor reproduce too much and the wealthy relatively too little, and that has often gone along with an assumption that the poor are also genetically inferior, so the human race is supposedly running down genetically.

Well, we see this a little differently now. The

truth of the matter is that it's precisely the well-to-do and the children of the well-to-do who make the most trouble. They're the biggest consumers and the biggest polluters. It's estimated now that an American child makes fifty times the demand on the world's resources that an Indian child does — and this takes not only a personal form but a national form. The United States, with 6 percent of the world's population, consumes about 40 percent of the world's irreplaceable resources and accounts for about 50 percent of the world's industrial pollution. So it is precisely the well-to-do, it is precisely the most developed nations, that are making the most trouble.

So, what can we do? For one thing, I think that, as quickly as possible, we have to see to it that convenient, safe, and cheap — I would much rather say free — means of contraception and abortion become universally available all over the world. I think the ideal we must strive to reach as quickly as possible is that nowhere in the world need a woman have an unwanted child. It is no favour to any child to be born unwanted. And there's something else that goes with what I'm saying, which is that all the time and everywhere, we have to take care of all the children, much better than we're doing now.

These problems are so overwhelming — they're coming on us so fast, they are so complex, they seem so impossibly difficult — that any sensible man recognizing these things would turn aside, would do something else, if there was something else to do. But that's just it: we have no alternative. We have to meet these problems; we have to solve them. We're not being asked; we are being told. It's life or death. So I don't feel like a prophet of doom; I feel like a prophet of hope. I am not downcast; I don't feel despair. I do sometimes feel a deep anger at having this great human enterprise degraded in this way for wholly trivial reasons. So I don't accept them, and I don't accept any such fate. I feel perfectly confident that we will solve those problems.

When I try to sum all this up for myself, to put it all in a nutshell, I want to tell you where I come out. We are all in such a hurry these days that none of us can afford to have a political philosophy that won't go on a button, and this one will go on a button. It is: *a better world for children*. As American children go, so goes America. As all children go, so goes humanity. When we've achieved a better world for children, that will be the better world for grown-ups too. But as we've seen, there are now far too many children, so I think we'd better add a word: a better world for fewer children — and then take good care of all of them, much better care than we do now.

That's my program for us, for you, for all humanity. Nowadays there is a beautiful passage from Deuteronomy always running through my thoughts. It's chapter 30, verse 19. Let me say it for you: "I have set before you life and death, blessing and curse; therefore choose life, that you and your descendants may live."

SIX

A QUESTION OF MEANING

HERE WE ARE, AT the end of these Massey Lectures, and most of all I just would like to talk to you. In the earlier lectures, I tried to structure rather carefully what I was saying, so as to teach and communicate and be as clear as possible. But that is behind us now, and I think the important thing is just to talk.

The questions I have raised — whence we come, what kind of thing we are, and at least a hint of what may become of us — those are essentially scientific questions, and I've tried to talk about them as a scientist, and tried never to stop, never to compromise with my science in talking about them. Why do I think it is so important to talk about them at all? Because it seems to me that those are the questions that have always needed to be answered, and need to be answered now, more than ever. I think that all our hope of going on now lies in the answers that we provide to those questions. I would even say that it's not necessary that the answers be right, or true, except in the sense that they guide us well.

Those are the questions men have always asked and tried to answer as best they could, and the answers to those questions have constituted their religions. That's what I think a religion is about: its function is to hold a society together and to prepare it to act or perhaps at times not to act. But at this time, in our society, surely it must be for action.

Among the many answers that have been offered traditionally to those questions, I think the ones that I have offered here are righter and truer than most. Indeed, they are as right and true as any that can now be obtained, and that's the special virtue, I hope, in having those questions asked and answered by a scientist.

I spoke of my own answers as constituting my religion, the wholly secular religion of one scientist. In doing so, I don't quarrel in any way with other more traditional systems of belief or thought, provided just one thing: that they keep us on the path of life. The Egyptian religion, for example, became obsessed with death. It came to serve death more than life, came to have cities of the dead that rivalled the cities of the living. The Aztecs got into similar trouble. They thought of their gods as malignant and always having to be fed human sacrifice. That became an obsession, so that they were constantly at war, and the whole purpose of war was not political domination so much as to capture victims for sacrifice.

Now nationalism as we know it in the present world has become such a religion of death. Charles de Gaulle called patriotism the love of one's country, but nationalism the hatred of other countries. Western civilization, under the aegis of Christianity, that religion of peace, has cultivated the technology of killing as has no other culture in human history. So what I have been saying, and saying as a scientist, is that we live in a reasonable, understandable universe, a universe that breeds life; life holds a high place in the universe, and man an especially high place in life; and as we come to understand these things, we begin to answer my three fundamental questions in a particularly reliable, credible, satisfying way.

That wholly secular religion of mine that contains no supernatural elements: does it leave much out? I'll be asking that question again shortly, but let me say now: not much; perhaps nothing of great value. Surely it doesn't leave out poetry, art, emotion, awe. And I'd like to say one more thing, which may surprise some of you: for me, it doesn't leave out the Bible. I'd like to talk about the Bible a little, because it is much more important to me than you might suppose. It is by far the most absorbing, exciting, and beautiful book I know. You've all heard that cliché: if you're going to a desert island and had one book, which would it be? That offers no problem for me: it would be the Bible. That isn't just a literary judgement. I don't think we have ever known a culture that hasn't a mythology, and that mythology constitutes its roots. The Bible holds my mythology; those are my roots.

A mythology is a frightfully important thing. We need those roots, and that's where they are. But much more than that, the Bible gives us a common imagery and metaphor we can all share, provided we know it. If we lose that, we've lost a lot. With that common imagery and metaphor, one can say a lot in a few words. And one more thing: the Bible gave us common speech — the best speech we've ever had. The King James Bible is the best English we've ever had. And once, up to a century ago, that was a common possession. You take such a thing as Abraham Lincoln's Gettysburg Address: "Four score and seven years ago, our fathers brought forth, on this continent, a new nation, conceived in Liberty, and dedicated to the proposition that all men are created equal." That kind of talk comes right out of the King James Bible. Anybody not brought up on the King James Bible wouldn't use words like that. They were once the words all Americans had at their command. They were once the words that all English-speaking people had at their command. So don't be surprised if I talk in terms of that Bible — I'm a scientist who reads the Bible.

Take the episode in the Garden of Eden. If that had been a Greek legend, the tree that yielded the forbidden fruit would have been the Tree of Knowledge. But it's a Jewish legend, so that tree is the Tree of the Knowledge of Good and Evil. This is the territory that we have been exploring, that strange province in which knowledge goes over into the knowledge of good and evil. We've come upon very difficult and painful times in which that distinction might mean everything.

I once heard the Danish physicist Niels Bohr say something in this province that meant very much to me. He was talking of the migrations of the eels — I talked of them in chapter 4, you may remember. All the eels of the shores of the Atlantic Ocean, European and American, come together in the depths of the Sargasso Sea to spawn; after spawning, the mature eels die and the little larval eels have to find their way back home alone, the American eels to America and the European eels to Europe. It takes a long time and we have no idea, we biologists, how they find their way back, but they do. Bohr was talking about that, and he said the unforgettable words "It's just because they do not know where they are going that they always do it perfectly." And that's it. We know, to a degree, where we are going. We've eaten of the fruit of that Tree of the Knowledge of Good and Evil, and in that knowledge have lost all assurance that we will do it perfectly. We're on our own. We have to choose our own way and can choose rightly or wrongly, for good or evil.

I have tried to tell you as a scientist that man is a great thing in the universe. In him, and in his like, matter comes to know itself and to understand itself. It's a kind of culmination. I think we've had enough of humble man. He has good reason to be proud, far too proud to be no more than a consumer, or an urban ant, or a faceless social or political unit, or anything less than a whole man.

There's been so much said of the nature of man that is quite different, but what I'm saying now I find in those opening paragraphs of the Book of Genesis, in that story about Adam and Eve in the garden. We've been told such wrong things about what that story is saying. It's spoken of as the fall of man and original sin, and we are told that's when God cursed man. But what was the curse? The answer is rather strange. The curse that God laid on man was to work for a living — not as terrifying as we are sometimes told. And I take great satisfaction in the end of the story. In that last encounter with Adam and Eve, what is God doing, that angry God? Well, it's a little surprising: he's making clothing for them. And what does he say at the last? He says, "Behold, the man has become as one of us." Get the plural: one of *us*, knowing good and evil. And now, that he not eat also of the Tree of Life and live forever, he has to leave the Garden. That is what God is saying in that story: that man has become God-like, lacking only immortality. And does he even lack immortality?

I've had to ask myself: If that wholly secular religion of the scientist is my religion, how good a religion is it? How does it compare with the traditional religions? What do I have to do without that the traditional religions offered? I think there are two things: first — and I am choosing my words carefully — science offers no encouragement for a belief in personal immortality. But you see, the more I think of personal immortality, the less I want it. I have the immortality I want. I've spoken of it earlier in these lectures: every creature now alive represents an unbroken line of life since life first arose on this planet, and three billion years in that unbroken line of life is represented in my germ plasm and your germ plasm, those eggs and sperm. They are our hold upon the past. They are our grip upon the future.

The second thing is that science offers no encouragement for the belief in some sentient intelligence that hears and answers prayers. That's no argument against praying. Praying may be a very useful thing. For one thing, in praying one formulates one's needs and hopes and desires, and formulating them is a step toward achieving them. I rather think that in praying, one is praying to oneself — and yet a special self, a higher self, all that one conceives that man may be. For most of us, that's as far as we can reach toward a concept of God. So, in my secular religion, those are two things I do without, but I don't feel them as a profound loss.

I must add that this wholly secular religion of a scientist offers us things, too. For one thing, it's full of acceptance: acceptance of reality. I do not know of anything better than reality, anything more awesome, more beautiful, offering us more hope. In many of the religions of the past, there has been so much rejection of quite ordinary and acceptable things. There have been such strange choices made. Religions have been full of terms of praise and disparagement, very curiously chosen: rejection of flesh, of matter, of mechanism, of chance. Chance is always referred to with other words attached, as "mere chance" and "nothing but chance," and rejected. What I have is an acceptance of reality, all reality. It's a yea-saying to the universe. And that's the way I want it.

A last thing that I think is very important indeed is in the nature of science and the nature of traditional religion. Science goes from question to question. Its job is to ask ever more searching and meaningful questions, and every answer is tentative — not necessarily wrong, but incomplete and needing further refinement and extension. As fast as we get those answers, they raise further questions, so that one has this constantly expanding, endless horizon that keeps stretching before us and will never be reached. The questions are ever so much more important than the answers, the answers are all tentative; and that's a profound difference between science and most traditional religions. Religions claim to provide answers — frequently final answers — and in those answers lies an enormous vulnerability. They are constantly defending things that have become rather indefensible and trying to hold positions that have become untenable. Often I ask myself, can religion, even in the traditional sense, be a religion of questions rather than answers, of seeking rather than finding? Wouldn't that be just as satisfactory, indeed perhaps more satisfactory, than those religions of answers?

I have heard the story that Gertrude Stein on her deathbed turned to the faithful Alice B. Toklas and said, "Alice, what is the answer?" and when Alice had no response, Gertrude is said to have asked, "Then, Alice, what is the question?"

I think we know the question. The question, the ultimate question, is one of meaning — the meaning of life, what it is all about. That's an endless question, and there is no answer. Yet it's a question we must forever keep on asking and trying to answer as best we can. My own answer is that the meaning

is in the endless unfolding of the universe — that
unfolding of the universe that in its good time has
bred life, that unfolding of life that in its good time
has bred men. And this raises the further questions: How is man to unfold? What lies ahead?

I have been asked those questions rather frequently. As a biologist, I've been asked, "Since you
think that man is still evolving, in what direction
is he evolving? What lies ahead in human evolution?" I try to tell people that the most wonderful
thing is that we don't know — we can't know. Natural selection takes its own directions, and they
are not to be read beforehand. What I can read is
pretty trivial. For example, as I mentioned before,
we've only lately lost our tails, and it seems to me
that we are in the process of losing our toes; they
aren't of any use anymore, they've become purely
decorative. Apart from those trivial things, how
is man to further unfold? As I see it, his hope lies
in becoming ever more human, ever more a man.
And what's that? Well, I think that brings us back
to those uniquely human attributes, the knowing
and the creating. Man is a uniquely science-making and art-making animal — that's what made him
a man, and I think therein lies his future.

Is that an elite idea? Are art and science only

for the few? It's true that when one thinks of professional scientists and artists, they are very few indeed. But I don't mean it that way. I think this idea of man takes us in a quite different direction, because every child is the beginnings of both a scientist and an artist. For the few who become professional scientists and artists, the wonderful thing is they need never grow up. They need never stop asking questions and trying to answer them, as all children do. They need never stop making and doing creative things, as all children do. They need never stop constantly trying to exceed themselves, as all children do — not exceed others, but exceed themselves. The scientist has the privilege of going on, always becoming the more and more learned small boy or small girl, the artist the more and more skillful small boy or small girl. There is nothing better.

The greatest men I have ever known in my life were Albert Einstein, Niels Bohr, and Alfred North Whitehead, and they were the freest men I have ever known, and the most childlike. They were, I hope you won't mind my saying, almost like small puppies, dashing into every corner, wherever something excited their curiosity, wherever it seemed as though there might be something of interest.

There were no fences around them. And so, when I ask myself what would be the touchstone of the good society, the best society, I think it is something more than calories, though I want the people to have the calories; something more than clothing and housing and all those vegetative things, though I want all of those things for all people. I want something more: I want those uniquely human things. I think the touchstone of that good society would not just be to feed and clothe and take care of people, but that it would be the society that was most productive of good science and good art. That would be the good society for children: constantly unfolding, constantly surprising, a busy world, always with new things to do and new things to learn; a world in which one need never cease to be a child; a world for knowing evermore, for creating evermore; a world in which that universe that brought us forth would find ever greater and greater fulfilment; and a better world thereby for all the children.

AN INTERVIEW WITH GEORGE WALD

THE ORIGINS OF LIFE

LEWIS AUERBACH: You said that all cells ferment. Does that mean that the cells in our body are fermenting? Do we ferment?

GEORGE WALD: We certainly do, and we do it on the grand scale. We respire — as indeed yeast does when it gets some oxygen — but if you were to, for example, start running fast, your respiration could keep up with that violent exercise for only the first few steps. You run a race not on respiration, but on fermentation, and the sign that you've been fermenting is when you stop the violent running, you're panting. You're breathing hard because you're still in the process of removing products of fermentation. But I have something sad to tell you about those products of our fermentation. When we ferment sugar, instead of making alcohol we make lactic acid. So by running fast, you can't get drunk; you can only get tired.

Muscular fatigue is just a piling up of lactic acid in the muscles and the blood, and as long as that acid is in you, it makes you breathe hard. As soon as you stop running, your respiration begins to catch up and burns off that lactic acid. And as soon as it's rid of it, you're back to normal and are breathing easily again. So that's the way it is. It's curious that we were cheated out of being able to ferment sugar to alcohol; it's just one enzyme that differentiates us from yeast in that regard. Yeast has it, we don't.

AUERBACH: Do you think that in one of these other possible worlds that you've talked about there might be animals like us that make alcohol?

WALD: That's an interesting question, and the answer is no. Because, let me say virtuously now, alcohol is a poison and yeast having made it has next to get away from it. If you let yeast perform an alcoholic fermentation in a closed space, at the end of the process you find the whole culture of yeast dead at the bottom of the bottle, because the yeast is poisoned by that alcohol.

As long as one is in a tremendous amount of water that can keep removing the alcohol as fast as it is made, one can afford to make it; but creatures like us who are much more self-contained and walk about on the land, and can't get rid of everything that we produce as by-products, excretory products, immediately, we'd be in a bad way — in fact, drunk most of the time — if we had the alcohol fermentation. So I think this is a good thing ultimately, having lactic acid as the product instead; one has the disability of fatigue, of growing tired, but we can cope with that more readily.

AUERBACH: One of the clearest implications one can draw from this talk is that where life might arise, it will arise. In other words, what is a probability must happen, and that's an amazing thing.

WALD: Let me say a word about that, because there is something so poorly understood, and that is the role of chance in the universe. Nothing can happen without there being a chance of its happening. Quite clearly, since we're here, there's a chance for this occurrence of life, and yet I speak of it as a highly improbable phenomenon, which is what it is. But, you see, what that small probability means is only that it takes more time.

That's the whole point: the things that are more probable happen sooner; the things that are less probable take more time. They're just as certain, really; they just take longer, and that's the way it is. You have to wait a long time for life to arise, but I think, given enough time, it's inevitable. Time is the hero of the plot, turning that very low probability eventually into certainty.

AUERBACH: Let's go on further for just a second. I have read a copy of your speech in which you talked about the probability of life ending on this planet, where you multiplied the 2 percent figure that a scientist had given you by fifty and came out with near certainty that there would, in fact, be some sort of total holocaust on this planet. I'll ask it another way: If there's a probability of life beginning, isn't there at least an equal probability of it ending?

WALD: I'll talk about those things a little later in this series, but let me say now that the speech you referred to involved the probability — not my assessment of it, but a so-called expert assessment — that we might have a full-scale nuclear war within the next period. That would mean life coming to an end because we had brought it to an end. Not a geological process, but a thoroughly human one, one that concerns us very much. That estimate came out because this expert said that he estimated the probability of full-scale nuclear war at something like 2 percent per year, and it's a very simple computation that would say that by 1990, there would be something like a chance of one in three, and by the year 2000, a chance of one in two, a 50 percent chance that we would have that full-scale nuclear war.

AUERBACH: If life arose, perhaps not so easily, but so certainly, why can't the scientist do it as successfully?

WALD: You're asking, "Why don't scientists, if they understand at least in principle how life arose, create life in the test tube?" This is something that people speak about a great deal, and occasionally we get news stories that it's happened, but it hasn't happened, and I think most non-biologists rather think that this is the big problem of our generation in biology, but not at all. There are very, very few biologists concerned with it, or directly and deeply interested in it. Very, very few biologists are working on or even near this problem.

I think that we can afford to be patient. We can wait awhile. My own opinion is that one of these days — I would estimate within the next twenty to thirty years — some biologist is going to bring together just the right things and in the right way, and we'll see the formation of a blob of living material, and he'll be pretty excited; and I doubt that I'll be around, but if I were, I'd be pretty excited too. That biologist will probably rush home and boast to his wife that he's just created life in the test tube, but the truth of the matter is that he wouldn't have created anything. He would have just set the conditions. He would have brought the right things together under the right circumstances, and then all he could do is watch; those molecules would organize themselves into that blob of living material.

THE ORIGINS OF MAN

AUERBACH: Listening to you, one might think that what you're saying is not science, but a sermon. Would you agree with that? Would you like to say something about that?

WALD: Well, yes, that's a very interesting thought and I'm glad you've raised it. You see, it's not a sermon at all, if by a sermon you mean the kind of thing that ministers ordinarily say to their flocks on Sunday mornings. It isn't that hortatory. It's not exhorting you. I'm not intentionally, in any case, using words to produce, perhaps reflexively, vague emotions — I do sometimes, I'm afraid, convey my own emotions, but that's a little different.

Nor am I using the authority of science to depart from it. That's been done all too often before. We all know those scientists who, having spoken with the authority of science, come to the point of saying, "Ah yes, but it isn't enough. There's something beyond, there's something else" — and with that, they go into another compartment and give us a religion that is distinct from science. Now, I'm trying to do something entirely different. To me, it's one world, one universe, and that's the whole point of what I'm saying. And what's the place of science in that universe? The place of science is to achieve understanding, and I'm not trying to do anything to an audience that isn't already done to me as a scientist, as an outcome of what I already know as science.

You see, I think mysticism is the easy way out. It's the lazy way to get to the same ends. The hard way is this long, painful business of careful observation and experiment, and the extraordinary thing is that that gets you there, too. And that's what these talks are about.

AUERBACH: If a religious viewpoint is a world view and science also entails some kind of world view, if these things are both true, then can't we at least in that sense say you are sermonizing on behalf of science?

WALD: Well, I started these talks by saying that what I was going to lay out was my wholly secular religion, the religion of one scientist, and that's what I'm trying to do here, but never compromising with my science. I began these talks by saying, men everywhere, at all times, have asked the same questions: whence we come, what kind of thing we are, and at least some hint of what may become of us. Whence we come — that's a scientific question, and I've been talking about it as a scientist. And what kind of thing we are, I think that's in the same universe, a scientific question, and I've been talking about that too. That hint of what may become of us, we still have one or two further talks in which to discuss that.

AUERBACH: You've said you would prefer biological specifications to technological specifications for society. I wonder what happens if democracy doesn't work, would you then favour a planned society, a technologically specified society?

WALD: Well, perhaps we'll get to talk about that too somewhat further, but let me answer you now. One of the troubles of this generation of young people is that my generation, their elders, tried an enormous variety of political experiments, particularly after World War I, and none of them, in my opinion, has proved very attractive. We've tried many kinds of planned society: communist, fascist, military dictatorships on the right, and now something somewhat new on the left. None of them seem to me to offer the hope for man, the humanity, the flexibility — indeed, the characteristics that I associate with living organisms — that democracy does. So I'll go on struggling to achieve that open-ended system in which one keeps experimenting and, as times change, one meets those changes in new ways and one does so because adaptation to conditions is built into the system; and that's my understanding of democracy.

AUERBACH: One can imagine, however, let's say a democratic socialism, socialism which involves planning and democracy which involves the open-endedness of which you talk. Is that possible within your view?

WALD: You understand, I'm not speaking for what we call anarchy. One always has plans, and there are always various factions in democracy in conflict with one another, and each of them has plans, each of them has a scheme of things. It's that struggle for existence that can lead to the outcome that we value. It's trying all these things out. One has a plan, but the plan is always an immediate plan, a short-range plan. I'm very suspicious of the long-range plans, because conditions constantly change, and however one devises the next scheme, the next thing to do, one is never sure of the outcome. It sometimes comes out very differently from how one has planned. So there's a great deal of having to just wait and see, and then adapt again, and that's the thing I have most confidence in.

One couldn't move at all without a program, without a plan, but that's the short-range game, and it's by no means clear that if one's plan is actually executed, put into effect, that it will have the consequences one had hoped for it; so there is all this reliance on feedback, just as in living organisms. One tries everything and then selects those things that work a little better, for their time and in their place, and there are no absolutes and there are no permanent solutions. That's the way it is with living organisms, and that's the way it would be with democracy working as it should.

ANSWERS

AUERBACH: I've noticed that all four of your previous lectures were more consciously an affirmation of the possibilities of life, of this goodness of life, of the awesomeness of the creative force, and here you seem, at least at the beginning of your lecture, to have departed very substantially from this affirmative kind of style. Why now? Have you lost your confidence?

WALD: Not at all. You remember in my very first lecture, I raised the problems of technology as opposed to science. I said, "Know all you can, but do only those things that are socially useful." Then I tried to really set the cosmic stage for men, to try to get my listeners to feel as I do, that he's a kind of culmination of a large part of the history of the universe, a great thing in the universe. Indeed, so great a thing, so noble a thing potentially, that he should behave as great and noble. In this fifth lecture, I've come to our times — our deeply troubled times. What I'm talking about in that little dinosaur parable is just going with that thought about technology and how necessary it is that the technology be pursued, not for the producers of technology, but for those who are going to live with the technology. So I don't think that this discussion is at all disconnected from what has gone before. It just brings us into the present time.

AUERBACH: That brings us to the dinosaur parable. Do you think that the dinosaur became extinct

because of the small size of his brain, that we therefore have a better chance?

WALD: I think that the whole point is that we've
become dinosaur-like. Those dinosaurs were very
early over-consumers, and in their own way polluters. They were just awfully big, much too big
for that amount of brain to carry. Now you ask the
question: Would a bigger brain have helped? Why
yes, because it would have helped them to catch up
with those enormous bodies by producing a technology that would have brought things into balance. They were out of balance. It was too much of
a problem to keep that much body going with that
small a brain.

We were about the right size, but then we began
expanding in the form of technology, our power,
and by now the proportion of power to brain —
the proportion of what was in dinosaurs brawn to
brain — has again grown to be enormous and has
brought us again to the brink of disaster. So we
again have to bring things back into balance.

AUERBACH: You say, "A better world for fewer children." And yet it seems, or at least some people say,
that the poorer countries, those countries with the

most children, are the ones that seem least likely to institute any form of birth control. How do you explain that? Is there any way around this apparent fact?

WALD: Well, I'm very glad you asked that question, because there is so much more to be said, and it is quite necessary to say. I believe that it's absolutely essential that we bring population under control. I believe that that's most essential precisely to the poor and the underdeveloped nations, and yet we find that not only those underdeveloped nations, but many of the poor in our own country, look on this call for universal birth control as a species of genocide. They reject it. They fight against it.

Now, why? I'd like to say why. When one becomes interested in population control and begins to work on it, one finds oneself with very strange bedfellows. For example, Chase National Bank is very much interested in birth control; as is Standard Oil of New Jersey—some of the most powerful financial and industrial forces in the country back up this call for birth control. Why? Well, for a rather strange reason that we'd better begin to understand. With the growth of technology, with the greater availability of machines to do the world's work, human muscle has ceased to be as important as it once was. We don't need as much human muscle as used to be useful in the world. There isn't nearly as much for it to do, and the possessors of those muscles, meanwhile, have been watching TV. They see how the affluent live. There is a revolution of expectation that's run all over the world. They want things. Their muscles aren't needed anymore, and yet they want food, housing, clothing, education, all kinds of the good things of life, so they've become rather an embarrassment, and for wrong reasons as well as for right, they're being offered, indeed urged to practise, birth control.

What's the solution for this difficulty? Well, I think the solution is to take care of people — to take care of all people. If you're taking care of them, you can talk to them this way. If you're not taking care of them, they have to take care of themselves, and take refuge in numbers, and that's what they're doing. So I think those two things go together: the call for limiting population with the responsibility for taking care of what people there are — and most of all for children, all children everywhere.

AUERBACH: Is it that when the standard of living

approaches subsistence or goes below subsistence that the desire, the need for procreation, for having children, becomes greater?

WALD: Well, there are certain automatic and pretty universal forces at work and let me say a word about them. It's perfectly true that all industrialized societies have considerably smaller birth rates than agrarian societies. Now as for need, that's a little more complicated. It is true that the more hands there are on the farm, the more work gets done — it's those human muscles. There are still places in the world, and farms are among them, where human muscles are useful. There is another factor, and that is that those agrarian societies frequently don't have the sanitation or the medical care that exists in developed countries, and so they tend to run very high infant mortality rates. One literally needs to start a big family in order to bring through even a reasonably sized one. So it is true that agrarian societies tend to run much larger birth rates than the industrialized.

There's another thing to say, and that's very, very important indeed, particularly for places such as China and India and Pakistan and Puerto Rico and others, and that is that throughout human history,

I don't think there has been a single instance of an agrarian society transforming itself into an industrial society without the population simultaneously doubling or tripling. That happens perfectly automatically. It isn't a matter of what people want. It's just the way these things go. You see, you go into some really poor and underdeveloped country and try to do something helpful, and you find that everything you do has the effect of decreasing infant mortality and, in fact, decreasing mortality in general. You dig a sewer, you bring in a doctor, you bring in more food — anything you do has this immediate effect that brings about a leap in population. Unless something strange happens that doesn't tend to happen and that never has happened before as you are doing your good to that country, as you are developing it, its population still tends to double or triple. And of course that could be a sheer disaster and be working against every hope of improving its condition.

AUERBACH: What do you say to the position of the Catholic Church against abortion and against contraception?

WALD: Well, of course that's an enormous problem,

and it worries many Catholics as well as non-Catholics. The Catholic position, particularly on abortion, is that it's a species of murder. I think that the essential question is: At what point in human life would an interference with conception constitute murder? I'm speaking as a biologist again: I think that any point one chooses is rather arbitrary. The point chosen by the Catholic Church, as I understand it, is the point at which the sperm entered the egg. Before that, one doesn't worry about it, and yet as a biologist, I look on those sperm cells as both alive and human, which they are. But we don't worry about them. The egg—every fertile woman passes an egg every twenty-eight days approximately—we don't worry about those eggs, but they are alive and human.

To me it's rather arbitrary to declare that a human life begins at the point at which the sperm enters the egg. Indeed, I know as a biologist that one can start an egg to becoming an embryo without the benefit of sperm. I think that the point at which an embryo becomes a person, that it achieves some degree of personality, is at birth. But this does raise very serious problems, and my feeling is that one of the things we most need, and it's being worked on very hard and with some prospect

of early success, is an abortion pill. The earlier it
works, the less trouble it's going to make psycho-
logically and theologically. I think that soon abortion will not be the difficult, expensive thing it is
now, and though not nearly as dangerous as child-
birth, it does carry certain dangers in the form of
surgery. I think soon we'll have an abortion pill and
all those things will be much easier.

I'd like to say one last thing: it's not as though
one were calling for abortion in a world not already
practising it. May I say that not only is abortion
perhaps the principal method of birth control in the
world today, but that it appears particularly prom-
inently in Catholic countries, precisely because
no other means of birth control are available.
And though the statistics on the extent of illegal
abortion are always suspect — it's a kind of statis-
tic that isn't easy to obtain — the best information
we have from the most reliable sources estimates
that in France and Italy at present there is about
one illegal abortion for each live birth. Throughout Latin America there seems to be about one illegal abortion for every two live births; from one
Latin American country to another, it goes up and
down. In Brazil, for example, there seems to be one
illegal abortion for every three live births, but in

Uruguay, we are told, there are three illegal abortions for each live birth.

It may interest you to know that in the United States, these estimates run about one illegal abortion for every four live births. Now it's also interesting to ask: "How does it go in countries in which abortion has been legalized?" Well, it's rather strange. The proportions are much the same as in those other countries where it's illegal, and indeed even in those Catholic countries for which I've been giving you the estimates. For example, in Japan, where abortion has been legal for a considerable time, there is about one legal abortion for each live birth, and in Hungary the statistics for 1964 are now 133,000 live births to 180,000 illegal abortions.

AUERBACH: Would you like to say something about the notion that a better world with fewer children is, in fact, something really profoundly conservative, something trying to preserve our way of life or at least approaching our present standard of living?

WALD: Well, it's highly conservative. That's what it's about. It's an attempt to preserve the quality of human life. And what does that mean? What is one trying to have for the children of this world? Is one trying to have an indolent life, in which they'll be served hand and foot and so on? Not at all.

I think what one would like for all the children of the Earth is to have them treated so that they would realize maximally, optimally, their genetic potentiality. All of us are born with a complement of genes that are, so to speak, permissive; they say what things are possible in us as individuals. Then, to actually realize those things, what does one need? Well, it really starts with decent prenatal care for mothers. The mothers need to be fed and then continue to be fed while they're nursing the children. Then one needs to give the children adequate food, clothing, housing, and a decent environment to grow up in, and eventually educate them to the extent of their needs and capacities.

It's a large order, but altogether doable, provided we bring population under control. And that's what one is talking about. We want a human race in which all children have a chance — a somewhat equalized chance — to realize their genetic potentiality, and then they're on their own. It will surely be a life that offers plenty of variety, plenty of competition. Some will go further and some will go less far. But all of them will have a chance — an almost equal chance, I would hope — to be what they might be potentially.

A QUESTION OF MEANING

AUERBACH: Well, the only thing I can think of now is to try and come down from these lofty heights and to find out from you what your present concerns are politically. I know you're not concerned about your own immortality, but you must be concerned about some things that are going on about you.

WALD: Yes, indeed, and that is the real point. What I've been talking about in these lectures is the base from which the immediate issues, the politics, emerge. I'm being asked all the time, "It's very nice to talk in such general terms, but do you have a program?" And you might ask me, "Does your program stop with such a thought as a better world for children?" No, it doesn't at all.

I have had to ask myself, of course, what the issues are, and by now I can recite them almost like a litany. I'm ever so much more interested in the issues than in the particular men who might represent them; indeed, I choose my men on the basis of promoting those issues. So what are the issues? Well, I'm an American, and some of them are American issues, but American issues have a way of affecting the whole world. So let me just list them for you without embellishment: I want us out of the Vietnam War. I want the draft to stop — a peacetime draft, a permanent draft is the most un-American thing I know. My parents came over here in steerage, and one of the things they were escaping was involuntary military service. It's a deep shock to me that this should have come to our country.

I want much smaller armed forces than we now have. I want a drastic cutback in the military budget, which will give us some money to satisfy human needs. So that starts a kind of positive program of issues: housing, much more housing, low-cost houses; schools and much more aid to education; public transportation, which has almost ceased to function in our country; clean air and clean water; much better nutrition — we now know officially through a presidential commission that something close to twenty million Americans are regularly hungry, including children who come to school without breakfast, and I'm a teacher and I know you can't teach a hungry child.

...ose are just a few of the issues. There's a ...itany of them and those are things I think are ...ll worth working on.

AUERBACH: How related do you think your pursuit of these issues is to the more general kinds of concerns that you've been talking about in these lectures?

WALD: I think one of the most important things now, one of the things we most lack, is some commonly accepted scheme of things. There'll never be just one scheme that's accepted by everyone; several will do, and preferably they will be rather compatible. That breeds the politics, so I think it's all very much connected. This is my own scheme of things, and the only reason for saying it in lectures such as these is there may be others who can accept it.

AUERBACH: Would you say that you spend more time on science now or more time on politics?

WALD: Well, that's a curious question, and I must admit with great regret that I'm spending a lot less time on science. I don't cut corners in my teaching;

I do all my teaching and I do it as thoroughly as I can. I've had to let myself be distracted to a great degree from my scientific work. I still have an active laboratory and people working in it who are doing lovely things, and that's a fine thing, that makes me very happy. But I don't get to do very much science myself, and I regret that very much. Every now and then I think, "Wouldn't it be better to do that next experiment rather than be distracted with these other things?" And then I feel such a sense of urgency — and I must say I feel it through my own children and through my students — that just now I think it's perhaps more important that I work on the politics, on trying to talk to people about the kinds of things that I've just finished talking over with you, than doing that next experiment. That may be a mistake, but I can't choose otherwise.

AUERBACH: I have the impression that you didn't emerge as a visible political figure until after this rather famous speech you gave, "A Generation in Search of a Future," in March 1969. I wonder if you could just tell us a little bit about the circumstances surrounding that speech and the kind of reaction you got to it.

WALD: Well, yes, I'd be glad to. Some students at MIT, who together with faculty organized the meeting, had asked me to speak. I'd been more and more aware that the students I teach at Harvard, who are mostly freshmen and sophomores — young people, young men and women — were increasingly upset, month by month. There were things that were deeply troubling them, that seemed to be making it harder and harder to teach them and harder and harder for them to learn, and of course I had to worry about that. I'd been asking myself and asking myself, what was wrong, what was bothering them? And of course it isn't just students at Harvard, or even just students in the United States — Canadian students, German students, Czechoslovakian students, Hungarian students, Mexican students, Spanish students — all over the world, both sides of the Iron Curtain, both sides of the Bamboo Curtain, young people are deeply upset, and there must be some reason for it.

I kept asking myself, "What reason?" And then the time came when I thought I knew the reason, and that's why I called that speech "A Generation in Search of a Future." That's a ghastly kind of thing to face and to say, but I'm glad I did, and my life has been very different since. I've come to know my

first major politicians, and I've been criss-crossing
the country and talking to all kinds of people —
not always people in universities, all kinds of people — and always trying to search for some way out
of these difficulties that will open up that future.

I still see no clarity; I don't think that we're making progress. I think that year by year and even
month by month, all those things that were bothering my students then have grown worse. But I don't
feel despair; I don't feel discouragement. Why?
Because I find the course of things and its inevitable
consequence wholly unacceptable. Having the view
I do of the nature of man and of the human enterprise, do you think that I can accept the thought
that it will all disintegrate now for wholly trivial
reasons? Because if it goes down now, it will be for
wholly trivial reasons: for reasons of profits and
power and status, all for just a few people. Is all
humanity and the human enterprise to be sacri-
ficed for that?

So I feel quite confident, though there are no
victories to show as yet. In a way, our problems
seem almost impossibly difficult to solve, and any
sensible man recognizing how complex and difficult they are would gladly turn aside and do something else. But there is no alternative. We're not

being asked, we're being told. There is no alternative, and so we'll do these things and we'll make that future for our children.

AUERBACH: You must have more than a faith that there will be a better future. What are you doing practically?

WALD: Well, you're raising the fundamental question that is raised in all religions, of faith and works. You've heard my faith; you want to know my works.

There's nothing to boast of, but I did start an enterprise lately that I'd like to tell you something about, because I have great hope in it. About two months ago, I wrote to Leonard Woodcock, the president of the United Auto Workers Union, saying that it seemed to me that if a channel were open for progressive elements in the labour unions to get together with progressive elements in the universities, both faculty and students, that that might be useful.

He wrote back and said it sounded good to him, would I set up a meeting. And so we had such a meeting about three weeks ago. Leonard Woodcock had invited top national executives from about eight very strong, very powerful unions, and they met with about three dozen faculty members from the Cambridge-Boston area and about fifteen students. We got along fine, so well that about two weeks later we had a second meeting. This time there were more unions, and we've decided to try and start a whole series of such meetings in key areas across the country, perhaps also in Canada, and have in each of those places a membership organization form itself, composed of union members and students and faculty members from the colleges and universities in the area. The whole strength of what we're trying to start would be in those local area groups. All we expect to have is a rather loosely structured national board, merely to coordinate and intercommunicate and perhaps handle national issues, such as bills before the Congress.

That's just the beginning, and I have high hope in what it may become and what it may do. It will give us in the universities what we most need now: a base in the outside community. It will give the working people whatever expertise and capacity for research and knowledge the universities might provide, a sort of a think tank, as one of the labour leaders called it, such as government and industry has possessed for many years, but that labour has never had; a disinterested think tank that can go to work on all kinds of problems of deep interest to working men, but equally frequently of interest to those working men who teach and study in the universities.

AUERBACH: It's interesting that at this stage of your life, you have tended to move somewhat away from science to political involvement. I wonder if you were to choose a career now whether you would choose science all over again?

WALD: I'm never sorry for having chosen science, for the reasons that I tried to lay out in these lectures. As a scientist, I feel I have my feet not only on the ground, but a little bit *in* the ground. I'm rooted in reality. That's very important to me. I think if you've got your feet on the ground, you can occasionally afford to have your head in the clouds. In any case, that's the whole basis from which I operate, and it's what I've been trying to tell you: it's my science that assures me that we're in a great and noble and worthwhile game that's cosmic in its magnitude, and I just won't see it go to pieces now. That gives me my job now — to see to it that the human enterprise goes on and continues to develop and continues to fulfill these wonderful potentialities.

INDEX

abortion, 76, 109–112
Adam and Eve, 85
adaptation, 102–103
agrarian societies, 108
alcohol, 24, 94–95
algae, 68
alienation, 5–6, 63
amino acids, 20
ammonia, 20, 21
amoebas, 45
art, 89–91
atmosphere, 20–22, 27, 67–68.
   See also environment
atomic energy, 11, 72–73
Auerbach, Lewis, xvii–xxii,
   93–123
automobiles, 64–65, 69

bees, 58
Beethoven, Ludwig van, 34
beryllium eight, 9
Bible, 42–43, 82–83
   Genesis, 16–17, 83, 85

birth control, 76, 106, 109–112
birth rates, 108
body (soma), 35, 51–52, 61–62
   death and, 52
   immortality of, 59–60, 61
Bohr, Niels, 83–84, 90–91

carbon, 9, 11, 21
carbon dioxide, 21, 24, 25, 28
carbon monoxide, 21
cars, 64–65, 69
Catholic Church, 109–112
chance, 87, 95–96
Chase National Bank, 106
chemistry, 19, 34–35
children. *See also* young people
   better world for, 77–78, 91,
      112–114
   caring for, 65, 107, 113–114,
      115
   deaths of, 108, 109
Chinese (ancient), 73
civil rights movement, xi

Cold War, 70–71
combustion (burning). *See also*
    respiration
  of gases, 7–8, 9
  of sugar, 26
Common Law (English), 40
competition, 36–37, 38
conjugation, 46–47
contraception, 76, 106, 109–112
creation
  in Bible, 16–17
  vs. production, 4–5
  of the universe, 6–11
creativity, 34–35, 89–91

Darwin, Charles, 36–37,
    39–40
DDT, 67
death, 45
  and body, 52
  cars as cause, 64–65
  natural, 60
  religion and, 80–81
  reproduction as precursor,
    50–55
de Gaulle, Charles, 81
democracy, 41, 101–102
The Descent of Man (Darwin),
    39–40
design
  organic, 36–39
  processes for, 36–37
  technological, 36, 38–39
Ding an sich (essence of
    things), 12–13
dinosaurs, 63–64, 65, 104–105
  humans as, 64–66, 105

disarmament, 73–74
division (reproduction by),
    45–50, 51, 52

Earth. *See also* life
  atmosphere of, 20–22, 27,
    67–68
  history, 12, 34
education, 115
eels, 53–54, 83–84
Einstein, Albert, 7–8, 90–91
electrons, 6, 12
elements (basic), 8–9, 11–12
  combining of, 20–22
  in organic molecules, 18–19
Enchytraeus fragmentosus
    (worm), 50
energy, 7–9
  of organisms, 23–24, 26
  sources, 11, 24, 26
  of stars, 9, 11, 25
English (language), 82–83
environment, 76, 115
  pollution of, 67–70
ethics, 3, 42–43
evolution, 23, 32, 39. *See also*
    natural selection
  of man, 43, 89–91
exertion (physical), 93–94, 95

fermentation, 23–25, 26, 93–95
fertility, 60, 108
flatworms, 48–50
fragmentation (reproduction
    by), 48–50
France, 111

galaxies, 30–31. *See also* Milky Way

Garden of Eden, 83, 85

Gauguin, Paul, 1

"A Generation in Search of a Future" (Wald), 66, 117–118

genetic material, 46, 51–52

germ plasm, 51–52, 61–62

Gettysburg Address (Lincoln), 83

God, 86. *See also* Bible; religion

Harvard University, 66, 118

Heisenberg Principle, 12

helium, 7, 8, 9

Hesse, Hermann, 3

housing, 115

humans. *See* man

Hungary, 112

hunger, 55–57, 115

hydras, 48

hydrogen, 7–8, 11, 20–22

immortality, 52, 59–60, 61–62, 86

industrialization, 69, 106–107, 108–109

infant mortality, 108, 109

inheritance, 37, 38, 46, 51–52

Italy, 111

Japan, 112

Johnston, David, xviii

Kant, Immanuel, 12–13

Labor-University Alliance, xiii–xiv, 120–122

lactic acid, 94, 95

Latin America, 111

law, 40

lemmings, 55–57

life (on Earth). *See also* molecules

  components needed, 11, 17–18

  first appearance, 23–24, 34, 110

  length of, 29–30, 61–62

  meaning of, 88–89

  origins, 16–18, 27–28, 30

  as part of nature, 17, 30, 32

  probability of, 30–31, 85–87

  quality of, 75, 112–114

  as supernatural creation, 16–17

  as technological product, 17, 43, 97–98

lightning, 22

light speed, 8

Lincoln, Abraham, 83

Luther, Martin, 42

mammals, 64, 65. *See also* man

man (humankind), 2

  alienation of, 5–6, 63

  destructive nature, 57, 63, 81

  as dinosaur, 64–66, 105

  evolution of, 43, 89–91

  first appearance, 23–24, 34, 64

  as molecules, 34–35

  nature of, 84–85, 89–91

man (humankind) (*continued*)
  troubles created, 67–76, 104
  and the universe, 13, 43–44
mass, 7–8
Medawar, Peter, 59–60
methane, 20, 21
migrations (mass), 53–57
  hunger as cause, 55–57
  reproduction as purpose, 53–55
military, 115. *See also* war
Milky Way, 30–31, 33–34
Miller, Stanley, 19–20
MIT (Massachusetts Institute of Technology), 66, 117–118
molecules (organic), 11, 12, 15–16. *See also* life
  geological processes and, 19–20
  living organisms and, 18–19
  photosynthesis and, 25–26
  in seawater, 22–23
  sources, 18–23, 25–26
mysticism, 100
mythology, 82. *See also* Bible

nationalism, 81
natural selection, 39–40. *See also* evolution
  at elemental level, 22–23
  in society, 40, 41
  as survival of the fittest, 36–37, 38, 57, 74
nature, 2, 16, 37
  life as part of, 17, 30, 32
neutrons, 6
Newton, Isaac, 34

nitrogen, 9, 11, 21
novas, 10
nuclear energy, 72–73
  and sunlight, 11, 72
nuclear war, 70–74, 97
nutrition, 115

oceans, 22–23, 68
oil industry, 68–69, 106
Oparin, Alexander, 22–23
organisms
  competition between, 36–37, 38
  multi-celled, 48–50
  single-celled, 45–48
  variations in, 37, 38
organization (states of), 33
oxygen, 9, 11, 21
  in atmosphere, 27, 28, 67–68
  life without, 23–24
  photosynthesis and, 25–26, 67–68
ozone, 27

Paramecium aurelia (protozoan), 46–47
patriotism, 81
photons, 6
photosynthesis, 28, 72
  and oxygen, 25–26, 67–68
Planaria (flatworms), 48–50
planets, 12, 30–31. *See also* Earth
Plato, 12–13
politics, 41, 114–117, 119, 122
pollution, 67–70

scientists and, 69–70
sources, 68–69
population size, 74–76,
 105–109
 industrialization and,
  106–107, 108–109
 and quality of life, 107–108,
  112–114
 reduction needed, 76–78,
  106
 wealth as factor, 75, 105–106,
  107
prayer, 86
prenatal care, 113
progress, 4, 41–42
protons, 7

quality of life, 75, 112–114. *See
 also* standard of living
questions (important), 79–80,
 87–91, 100–101
 answers to, 80, 87–88,
  103–114
 religion and, 16–17, 88
 science and, 87–88, 89–91

radiation, 6, 7–8, 21
 ultraviolet, 22
reality, 13, 122
 science and, 3, 87–88, 100,
  122
 understanding, 3, 87
regeneration, 49–50
religion, 86, 87. *See also* Bible
 answers provided, 16–17, 88
 and death, 80–81
 science and, 2, 99–100

reproduction. *See also*
 population size
 by conjugation, 46–47
 as death precursor, 50–55
 by division, 45–50, 51, 52
 as fitness measure, 74
 by fragmentation, 48–50
 as irrelevant, 58–59
 man and, 57–58
 migrations and, 53–55
 sexual, 50–55
respiration, 26, 28, 92, 94
Russell, Richard, 71

salmon, 52–53
science. *See also* scientists
 ethics of, 3, 42
 purpose, 5–6, 16
 questions asked by, 87–88,
  89–91
 and reality, 3, 87–88, 100,
  122
 and religion, 2, 99–100
 as religion, 2, 80, 81–82,
  85–87, 100–101
 and technology, 4–5
scientists
 as life-creators, 17, 97–98
 and pollution, 69–70
 as question-askers, 89–91,
  100–101
sea anemones, 48
Shakespeare, William, 34–35
Siddhartha (Hesse), 3
socialism, 102
society, 2, 58–59. *See also*
 population size

society (*continued*)
  art and science in, 91
  planned, 101–103
  and reproduction, 108
solar wind, 21
soma. *See* body
Soviet Union, 70–71. *See also*
  nuclear war
standard of living, 107–108.
  See also quality of life
Standard Oil of New Jersey, 106
Starbird, General Alfred,
  70–71
stars, 6–12. *See also* sun;
    universe
  death of, 8–10
  first-generation, 6–8, 11
  later-generation, 10–11
  lifespan, 29–30
  main sequence (maturity
    period), 8, 29
Stein, Gertrude, 88
Stoppenbrink, F., 49–50
sugar, 24, 25–26
suicide (mass), 55–57
sun, 8, 20, 29, 34. *See also* stars
  light from, 11, 72
supernovas, 10
survival of the fittest. *See*
  natural selection

technology
  dangers of, 105
  and design, 36, 38–39
  of killing, 70–74, 81
  life as product of, 17, 43,
    97–98

use for good, 4–5, 104
  and work, 106–107
Ten Commandments, 42–43
think tanks, 121–122
Toklas, Alice B., 88
tolerance, 41
transportation, 64–65, 115
Tree of Knowledge, 83, 84
troubles, 67–76, 104
  dealing with, 77, 118–122
  effects on young people, 66,
    101, 118
  environmental pollution,
    67–70
  nuclear threat, 70–74, 97
  overpopulation, 74–76,
    105–109

uniformity, 41
The Uniqueness of the Individual
  (Medawar), 59–60
United Auto Workers Union,
  120–121
United States
  abortion rates, 112
  automobile deaths, 64–65
  Constitution, 40
  resource demands, 76
  vs. Soviet Union, 70–71
universe. *See also* stars
  chance and, 95–96
  creation of, 6–11
  man and, 13, 43–44
  and meaning of life, 89
  states of organization of, 33
universities, 66, 117–118,
  120–122

urea, 19
Urey, Harold, 21

variation, 37, 38, 41–42
Vietnam War, xi, xix, 115

Wald, George, ix–xv, xvii–xxii
  interview with, 93–123
  MIT speech (1969), 66,
    117–118
  and politics, 114–117, 119
  religion of, 2, 80, 81–82,
    85–87, 100–101
  as scientist, 117, 122–123
  as teacher, 116–117
war
  nuclear (threat of), 70–74, 97

technology of, 70–74, 81
  in Vietnam, xi, xix, 115
water, 20, 21. *See also* oceans
  renewal of, 28–29
  salt, 18
Weismann, August, 50–52
Whitehead, Alfred North,
  89–90
Wöhler, Friedrich, 19
Woodcock, Leonard, 120–121
Woodruff, Lorande, 46–47
worms, 46–48

yeast, 24, 94–95
young people (1960s), 66, 101,
  118

(THE CBC MASSEY LECTURES SERIES)

In Search of a Better World
Payam Akhavan
978-1-4870-0200-8 (p)

The Return of History
Jennifer Welsh
978-1-4870-0242-8 (p)

History's People
Margaret MacMillan
978-1-4870-0137-7 (p)

Belonging
Adrienne Clarkson
978-1-77089-837-0 (p)

Blood
Lawrence Hill
978-1-77089-322-1 (p)

The Universe Within
Neil Turok
978-1-77089-015-2 (p)

Winter
Adam Gopnik
978-0-88784-974-9 (p)

Player One
Douglas Coupland
978-0-88784-972-5 (p)

The Wayfinders
Wade Davis
978-0-88784-842-1 (p)

Payback
Margaret Atwood
978-0-88784-810-0 (p)

The City of Words
Alberto Manguel
978-0-88784-763-9 (p)

The Lost Massey Lectures
Bernie Lucht, ed.
978-0-88784-217-7 (p)

More Lost Massey Lectures
Bernie Lucht, ed.
978-0-88784-801-8 (p)

The Ethical Imagination
Margaret Somerville
978-0-88784-747-9 (p)

Race Against Time
Stephen Lewis
978-0-88784-753-0 (p)

A Short History of Progress
Ronald Wright
978-0-88784-706-6 (p)

The Truth About Stories
Thomas King
978-0-88784-696-0 (p)

Beyond Fate
Margaret Visser
978-0-88784-679-3 (p)

The Cult of Efficiency
Janice Gross Stein
978-0-88784-678-6 (p)

The Rights Revolution
Michael Ignatieff
978-0-88784-762-2 (p)

The Triumph of Narrative
Robert Fulford
978-0-88784-645-8 (p)

Becoming Human
Jean Vanier
978-0-88784-809-4 (p)

The Elsewhere Community
Hugh Kenner
978-0-88784-607-6 (p)

The Unconscious Civilization
John Ralston Saul
978-0-88784-731-8 (p)

On the Eve of the Millennium
Conor Cruise O'Brien
978-0-88784-559-8 (p)

Democracy on Trial
Jean Bethke Elshtain
978-0-88784-545-1 (p)

Twenty-First Century Capitalism
Robert Heilbroner
978-0-88784-534-5 (p)

The Malaise of Modernity
Charles Taylor
978-0-88784-520-8 (p)

Biology as Ideology
R. C. Lewontin
978-0-88784-518-5 (p)

The Real World of Technology
Ursula Franklin
978-0-88784-636-6 (p)

Necessary Illusions
Noam Chomsky
978-0-88784-574-1 (p)

Compassion and Solidarity
Gregory Baum
978-0-88784-532-1 (p)

Prisons We Choose to Live Inside
Doris Lessing
978-0-88784-521-5 (p)

Latin America
Carlos Fuentes
978-0-88784-665-6 (p)

Nostalgia for the Absolute
George Steiner
978-0-88784-594-9 (p)

Designing Freedom
Stafford Beer
978-0-88784-547-5 (p)

The Politics of the Family
R. D. Laing
978-0-88784-546-8 (p)

The Real World of Democracy
C. B. Macpherson
978-0-88784-530-7 (p)

The Educated Imagination
Northrop Frye
978-0-88784-598-7 (p)

Available in fine bookstores and at www.houseofanansi.com

LOVE THE MASSEY LECTURES? THERE'S AN APP FOR THAT!

Available for free on the iTunes App Store, the award-winning Massey Lectures iPad App immerses users in the Massey universe. Winner of the Silver Cannes Lion award for Digital Online Design, the app brings together, for the first time, the full text of the CBC Massey Lectures with the audio recordings of the live lectures, and contains free bonus content, including discussions, related articles, video interviews, and more.

Uncover the full legacy and history of the CBC Massey Lectures Series, from 1961 to today. Learn more about selected Massey authors, their lives, their achievements, and their beliefs. Explore the complex web of themes within the Massey universe, and hear unique thoughts and insights from the lecturers. And contribute to the conversation yourself.

Download at bit.ly/MasseysApp

The Massey Lectures iPad App was conceived, designed, and developed by Critical Mass, a global digital marketing agency.